Differential Equations in Abstract Spaces

This is Volume 85 in
MATHEMATICS IN SCIENCE AND ENGINEERING
A series of monographs and textbooks
Edited by RICHARD BELLMAN, *University of Southern California*

The complete listing of the books in this series is available from the Publisher
upon request.

Differential Equations in Abstract Spaces

G. E. LADAS and V. LAKSHMIKANTHAM

Department of Mathematics
University of Rhode Island
Kingston, Rhode Island

 *1972*

ACADEMIC PRESS   New York and London

ACADEMIC PRESS, INC.
111 Fifth Avenue, New York, New York 10003

United Kingdom Edition published by
ACADEMIC PRESS, INC. (LONDON) LTD.
24/28 Oval Road, London NW1 7DD

LIBRARY OF CONGRESS CATALOG CARD NUMBER: 72-182660

AMS (MOS) 1970 Subject Classifications: 34G05, 47H15

PRINTED IN THE UNITED STATES OF AMERICA

Contents

Preface   ix
Acknowledgments   xi

CHAPTER 1   **The Calculus of Abstract Functions**

1.0 Introduction   1
1.1 Abstract Functions   2
1.2 The Mean Value Theorem   3
1.3 The Riemann Integral Abstract Functions   4
1.4 Abstract Lebesgue Integrals   8
1.5 The Abstract Stieltjes Integral   10
1.6 Gateaux and Fréchet Differentials   12
1.7 Notes   20

CHAPTER 2   **Semigroups of Operators**

2.0 Introduction   21
2.1 Strongly Continuous Semigroups of Operators   23
2.2 The Infinitesimal Generator   26

2.3   The Hille–Yosida–Phillips Theorem   32
2.4   Linear Autonomous Funtional Differential Equations   42
2.5   Analytic Semigroups   48
2.6   Notes   54

CHAPTER 3   **Linear Evolution Equations of the Parabolic**
            Type: Sobolevski–Tanabe Theory

3.0   Introduction   55
3.1   Definitions and Hypotheses   56
3.2   Statements of the Main Theorems and Some Heuristic Arguments   60
3.3   Properties of the Semigroup $\{\exp[-tA(\tau)]\}$ 62
3.4   Existence of a Fundamental Solution   74
3.5   Uniqueness of the Fundamental Solution   79
3.6   Solution of the Abstract Cauchy Problem   84
3.7   Differentiability of Solutions   88
3.8   Asymptotic Behavior   91
3.9   Notes   93

CHAPTER 4   **Evolution Inequalities**

4.0   Introduction   94
4.1   Lower Bounds, Uniqueness, and Convexity (Special Results)   95
4.2   Lower Bounds, Uniqueness, and Convexity (General Results)   106
4.3   Approximate Solutions, Bounds, and Uniqueness   118
4.4   Application to Parabolic Equations   122
4.5   Notes   125

CHAPTER 5   **Nonlinear Differential Equations**

5.0   Introduction   126
5.1   Counterexamples   127
5.2   Existence and Uniqueness   133
5.3   Nonlinear Variation of Constants Formula   141
5.4   Stability and Asymptotic Behavior   152
5.5   Chaplygin's Method   157
5.6   Global Existence and Asymptotic Equilibrium   161
5.7   Lyapunov Functions and Stability Criteria   167
5.8   Notes   171

CHAPTER 6   **Special Topics**

6.0   Introduction   172
6.1   Nonlinear Semigroups and Differential Equations   173

6.2   Functional Differential Equations in Banach Spaces   185
6.3   Second-Order Evolution Equations   191
6.4   Notes   199

Appendixes   200

Bibliography   211

Index   215

Preface

The theory of differential equations in abstract spaces is a fascinating field with important applications to a number of areas of analysis and other branches of mathematics. At the present time, there is no single book that is self-contained and simple enough to appeal to the beginner. Furthermore, if one desires to give a course so as to expose the student to this branch of research, such a book becomes handy. This being the motivation, the aim of our book is, in fact, to introduce the nonspecialist to this elegant theory and powerful techniques. But for some familiarity with the elements of functional analysis, all the important results used in this book are carefully stated in the appendixes so that, for the most part, no other references are needed. The required theory, from the calculus of abstract functions and the theory of semigroups of operators, used in connection with differential equations in Banach spaces is treated in detail.

We have tried to present the fundamental theory of differential equations in Banach spaces: the first three chapters form an integrated whole together with, perhaps, Sections 6.1 and 6.3 of Chapter 6. Chapter 4 is devoted to the study of differential inequalities, mostly, in Hilbert spaces. The theory developed in Chapter 5 is interesting in itself and could be read independently. This also applies to Chapter 4. Throughout the book we give a number of examples and applications to functional and partial differential equations which help to illustrate the abstract results developed. In most sections there are several problems with hints directly related to the material in the text. The notes at the end of each chapter indicate the sources which have been consulted and those whose ideas are developed. Several references are also included for further study on the subject. We hope that the reader who is familiar with the contents of this book will be fully equipped to contribute to this field as well as read with ease the current literature.

Acknowledgments

We wish to express our warmest thanks to Professor Richard Bellman whose interest and enthusiastic support made this work possible. We are immensely pleased that our book appears in his series.

This work is the outgrowth of lecture notes that were used in a graduate course at the University of Rhode Island. We are thankful to our colleagues E. Roxin, R. Driver, and T. Hallam and our graduate students, especially, M. Berman, G. Ladde, and D. Wood for their enthusiastic participation and constructive criticisms which helped to improve the exposition. We finally express our appreciation to Mrs. Rosalind Shumate for her excellent typing of the manuscript.

Chapter 1

The Calculus of Abstract Functions

1.0. Introduction

In this preliminary chapter the reader will be familiarized with those parts of the calculus of abstract functions that are essential in the study of differential equations in Banach and Hilbert spaces. By an abstract function we mean a function mapping an interval of the real line into a Banach space. We begin by defining weak and strong continuity and differentiability of abstract functions and prove a form of the mean value theorem for abstract functions. Next we develop the Riemann integral for abstract functions and those properties of this integral which are constantly used in the text. We then outline abstract integrals of the Lebesgue type (Pettis and Bochner integrals) and state some basic results. We also sketch the abstract Stieltjes integral for functions mapping a Banach space into another Banach space. Finally we treat in some detail the Gateaux and Fréchet differential of functions mapping a Banach space into another Banach space.

1.1. Abstract Functions

Let X be a Banach space over the field of real numbers and for any $x \in X$, let $\|x\|$ denote the norm of x. Let J be any interval of the real line R. A function $x : J \to X$ is called an *abstract* function. A function $x(t)$ is said to be *continuous* at the point $t_0 \in J$, if $\lim_{t \to t_0} \|x(t) - x(t_0)\| = 0$; if $x : J \to X$ is continuous at each point of J, then we say that x is continuous on J and we write $x \in C[J, X]$.

Abstract functions are in many ways reminiscent of ordinary functions. For example, a continuous abstract function maps compact sets into compact sets. Also, a continuous abstract function on a compact set is uniformly continuous. These statements can be proved in the same way that we prove them in a metric space.

An abstract function $x(t)$ is said to be

(i)   *Lipschitz continuous* on J with Lipschitz constant K if

$$\|x(t_1) - x(t_2)\| \leqslant K|t_1 - t_2|, \qquad t_1, t_2 \in J;$$

(ii)   *uniformly Hölder continuous* on J with Hölder constant K and exponent β, if

$$\|x(t_1) - x(t_2)\| \leqslant K|t_1 - t_2|^\beta, \qquad t_1, t_2 \in J, \quad 0 < \beta \leqslant 1.$$

It is clear that Lipschitz continuity implies Hölder continuity (with $\beta = 1$) but the converse fails as the classical example $x(t) = \sqrt{t}$, $K = 1$, $\beta = \frac{1}{2}$, shows. The (strong) derivative of $x(t)$ is defined by

$$x'(t) = \lim_{\Delta t \to 0} [x(t + \Delta t) - x(t)]/\Delta t$$

where the limit is taken in the strong sense, that is,

$$\lim_{\Delta t \to 0} \|[x(t + h) - x(t)]/h - x'(t)\| = 0.$$

The foregoing concepts of continuity and differentiability are defined in the *strong* sense. The corresponding *weak* concepts are defined as follows. Let X^* denote the conjugate of X, that is, the space of all bounded linear functionals on X. An abstract function $x(t)$ is said to be *weakly continuous* (*weakly differentiable*) at $t = t_0$ if for every $\phi \in X^*$, the scalar function $\phi[x(t)]$ is continuous (differentiable) at $t = t_0$. In the sequel, limits shall be understood in the strong sense unless we write w-lim to indicate that we are taking the weak limit. Also continuity and differentiability shall be understood in the strong sense unless otherwise specified.

A family $F = \{x(t)\}$ of abstract functions with domain $[a, b]$ and range in X is said to be *equicontinuous* if for every $\varepsilon > 0$ there exists a $\delta = \delta(\varepsilon)$ which depends only on ε such that for every $t_1, t_2 \in [a, b]$ with $|t_1 - t_2| < \delta$, $|x(t_1) - x(t_2)| < \varepsilon$ for all $x \in F$.

The following form of the Ascoli–Arzela theorem for abstract functions will be useful. Its proof is a special case of a more general theorem [63].

THEOREM 1.1.1. Let $F = \{x(t)\}$ be an equicontinuous family of functions from $[a, b]$ into X. Let $\{x_n(t)\}_{n=1}^{\infty}$ be a sequence in F such that for each $t_1 \in [a, b]$ the set $\{x_n(t_1) : n \geqslant 1\}$ is relatively compact in X. Then, there is a subsequence $\{x_{n_k}(t)\}_{k=1}^{\infty}$ which is uniformly convergent on $[a, b]$.

1.2. The Mean Value Theorem

For real-valued functions $x(t)$, the mean value theorem is written as an equality

$$x(b) - x(a) = x'(c)(b-a), \qquad a < c < b.$$

There is nothing similar to it as soon as $x(t)$ is a vector-valued function as one can see from the example $x(t) = (-1 + \cos t, \sin t)$ with $a = 0$ and $b = 2\pi$.

For abstract functions the following form of the mean value theorem is useful.

THEOREM 1.2.1. If $x \in C[[a, b], X]$ and $\|x'(t)\| \leqslant K$, $a < t < b$, then

$$\|x(b) - x(a)\| \leqslant K(b-a).$$

Proof: Choose a functional $\phi \in X^*$ such that $\|\phi\| = 1$ and $\phi[x(b) - x(a)] = \|x(b) - x(a)\|$. Such a choice of ϕ is possible in view of Appendix III. Define the real-valued function $f(t) = \phi[x(t)]$. Then

$$[f(t+h) - f(t)]/h = \phi[x(t+h) - x(t)]/h.$$

Since ϕ is a continuous linear functional and $x'(t)$ exists, it follows that $f'(t)$ exists for $a < t < b$ and $f'(t) = \phi[x'(t)]$. Hence, the classical mean value theorem applies to $f(t)$ and consequently there exists a τ, such that

$$f(b) - f(a) = f'(\tau)(b-a), \qquad a < \tau < b. \tag{1.2.1}$$

In view of (1.2.1) and the choice of ϕ we obtain

$$\|x(b) - x(a)\| = \phi[x(b) - x(a)] = f'(\tau)(b-a) = \phi[x'(\tau)](b-a)$$
$$\leqslant \|x'(\tau)\|(b-a) \leqslant K(b-a)$$

and the proof is complete.

COROLLARY 1.2.1.   If $x \in C[[a,b], X]$ and $x'(t) = 0$, $a < t < b$, then $x(t) \equiv \text{const}$.

PROBLEM 1.2.1.   Let $x \in C[[a,b], X]$ and $f \in C[[a,b], R]$. Assume that x and f have derivatives on $[a,b] - D$ where D is a denumerable set and $\|x'(t)\| \leqslant f'(t), t \in [a,b] - D$. Then

$$\|x(b) - x(a)\| \leqslant f(b) - f(a).$$

1.3. The Riemann Integral for Abstract Functions

Here we shall define the Riemann integral for abstract functions and prove the fundamental theorem of calculus. We also define improper integrals and discuss some properties which will be constantly used in this book.

Let $x : [a,b] \to X$ be an abstract function. We denote the partition $(a = t_0 < t_1 < \cdots < t_n = b)$ together with the points τ_i $(t_i \leqslant \tau_i \leqslant t_{i+1}, i = 0, 1, 2, \ldots, n-1)$ by π and set $|\pi| = \max_i |t_{i+1} - t_i|$. We form the Riemann sum

$$S_\pi = \sum_{i=0}^{n-1} (t_{i+1} - t_i) x(\tau_i).$$

If $\lim S_\pi$ exists as $|\pi| \to 0$ and defines an element I in X which is independent of π, then I is called the *Riemann integral* of the function $x(t)$ and is denoted by

$$I = \int_a^b x(t)\, dt.$$

THEOREM 1.3.1.   If $x \in C[[a,b], X]$, then the Riemann integral $\int_a^b x(t)\, dt$ exists.

The proof of this theorem makes use of the facts that a continuous function on a closed, bounded interval is uniformly continuous and that X is complete. We shall omit the proof.

Using the definition of Riemann integral one can easily verify the following properties:

(i)
$$\int_a^b x(t)\, dt = -\int_b^a x(t)\, dt$$

provided that one of the integrals exist.

(ii)
$$\int_a^b x(t)\, dt = \int_a^c x(t)\, dt + \int_c^b x(t)\, dt, \qquad a < c < b$$

provided that the integral on the left exists.

(iii)   If $x(t) \equiv x_0$ for all $t \in [a, b]$, then

$$\int_a^b x_0\, dt = (b-a)\, x_0.$$

(iv)   If $t = \omega(\tau)$ is an increasing continuous function on $[\alpha, \beta]$ with $a = \omega(\alpha)$ and $b = \omega(\beta)$, then

$$\int_a^b x(t)\, dt = \int_\alpha^\beta x[\omega(\tau)]\, \omega'(\tau)\, d\tau$$

provided that the integral on the left exists.

(v)   If $x \in C[[a, b], X]$, then

$$\left\| \int_a^b x(t)\, dt \right\| \leq \int_a^b \|x(t)\|\, dt.$$

Indeed, from the definition of the Riemann sum we have

$$\|S_\pi\| \leq \left\| \sum_{i=0}^{n-1} (t_{i+1} - t_i) x(\tau_i) \right\|$$

$$\leq \sum_{i=0}^{n-1} (t_{i+1} - t_i)\, \|x(\tau_i)\|$$

and the result follows by taking limits as $|\pi| \to 0$ and the fact that $\|x(t)\|$ is continuous and hence integrable on $[a, b]$.

THEOREM 1.3.2.   If $\{x_n(t)\}$ is a sequence of continuous abstract functions which converges uniformly to a necessarily continuous, abstract function $x(t)$ on the interval $[a, b]$, then

$$\lim_{n \to \infty} \int_a^b x_n(t)\, dt = \int_a^b x(t)\, dt.$$

Proof: We have

$$\left\| \int_a^b x_n(t)\, dt - \int_a^b x(t)\, dt \right\| \leqslant \int_a^b \|x_n(t) - x(t)\|\, dt$$

$$\leqslant \max_{[a,b]} \|x_n(t) - x(t)\|\, (b-a) \to 0 \qquad \text{as } n \to \infty,$$

which proves the stated result.

THEOREM 1.3.3.   If $x \in C[[a, b], X]$, then

$$(d/dt) \int_a^t x(s)\, ds = x(t), \qquad a \leqslant t \leqslant b.$$

Proof: Set $y(t) = \int_a^t x(s)\, ds$. Then, in view of the fact that $x(t)$ is uniformly continuous on $[a, b]$, we have

$$\|[y(t+h) - y(t)]/h - x(t)\| = \left\| h^{-1} \int_t^{t+h} [x(s) - x(t)]\, ds \right\|$$

$$\leqslant \max_{|s-t| \leqslant |h|} \|x(s) - x(t)\| \to 0 \qquad \text{as } h \to 0$$

and the proof is complete.

THEOREM 1.3.4.   If the function $x: [a, b] \to X$ is continuously differentiable on (a, b), then for any $\alpha, \beta \in (a, b)$ the following formula is true:

$$\int_\alpha^\beta x'(s)\, ds = x(\beta) - x(\alpha).$$

Proof: By Theorem 1.3.3

$$(d/dt)\left[\int_\alpha^t x'(s)\, ds - x(t) \right] \equiv 0, \qquad \alpha \leqslant t \leqslant \beta.$$

Hence

$$\int_\alpha^t x'(s)\, ds - x(t) = \text{const.} \tag{1.3.1}$$

For $t = \alpha$ we find the value of the const $= -x(\alpha)$ and the result follows by setting $t = \beta$ in (1.3.1).

REMARK 1.3.1.   In elementary calculus, if x is continuous on $[a, b]$, then

$$\int_a^b x(t)\, dt = (b-a)x(\xi)$$

for some $\xi \in (a, b)$. This is not true for vector-valued continuous functions x as we can see from the simple example $x(t) = (\cos t, \sin t)$, $a = 0$ and $b = \pi$.

Let $x: [a, b) \to X$ be an abstract function which is not defined at $b \leqslant \infty$. The *improper integral* $\int_a^b x(t) \, dt$ is defined as

$$\lim_{\varepsilon \to 0^+} \int_a^{b-\varepsilon} x(t) \, dt \qquad \text{if} \quad b < \infty$$

and as

$$\lim_{M \to \infty} \int_a^M x(t) \, dt \qquad \text{if} \quad b = \infty$$

provided that the limit exists.

The following theorem which asserts that integration commutes with closed operators (in particular, integration commutes with bounded operators) will be used often.

THEOREM 1.3.5.   Let A on $D(A)$ be a closed operator in the Banach space X and $x \in C[[a, b), X]$ with $b \leqslant \infty$. Suppose that $x(t) \in D(A)$, $Ax(t)$ is continuous on $[a, b)$ and that the improper integrals

$$\int_a^b x(t) \, dt \qquad \text{and} \qquad \int_a^b Ax(t) \, dt$$

exist. Then

$$\int_a^b x(t) \, dt \in D(A) \qquad \text{and} \qquad A \int_a^b x(t) \, dt = \int_a^b Ax(t) \, dt.$$

Proof: We shall prove the theorem when $b < \infty$. The case $b = \infty$ is left to the reader. Set $c = b - \varepsilon$ where $\varepsilon > 0$ is sufficiently small. For a partition π of $[a, b]$ we have

$$f_n \equiv \sum_{i=0}^{n-1} x(\tau_i)(t_{i+1} - t_i) \in D(A)$$

and

$$g_n \equiv \sum_{i=0}^{n-1} Ax(\tau_i)(t_{i+1} - t_i) = Af_n.$$

In view of the hypotheses, as $n \to \infty$ and $|\pi| \to 0$

$$f_n \to \int_a^c x(t) \, dt \qquad \text{and} \qquad Af_n \to \int_a^c Ax(t) \, dt.$$

Since A is a closed operator on $D(A)$, it follows that

$$\int_a^c x(t) \, dt \in D(A) \qquad \text{and} \qquad A \int_a^c x(t) \, dt = \int_a^c A x(t) \, dt.$$

Setting $c = b - n^{-1}$ in the previous result and using the definition of an improper integral and the fact that A is closed, the desired result follows upon taking limits as $n \to \infty$.

PROBLEM 1.3.1.   Define the rectangle

$$R_0 = \{(t, x) \in R \times X : |t - t_0| \leqslant \alpha, \quad \|x - x_0\| \leqslant \beta\}.$$

Let $f : R_0 \to X$ be a function continuous in t for each fixed x

$$\|f(t, x)\| \leqslant M, \qquad (t, x) \in R_0$$

and

$$\|f(t, x_1) - f(t, x_2)\| \leqslant K \|x_1 - x_2\|, \qquad (t, x_1), (t, x_2) \in R_0.$$

Let α, β, K, M be positive constants such that $\alpha M \leqslant \beta$. Then there exists one and only one (strongly) continuously differentiable function $x(t)$ such that

$$dx(t)/dt = f[t, x(t)], \qquad |t - t_0| \leqslant \alpha \qquad \text{and} \qquad x(t_0) = x_0.$$

[Hint: Use Theorems 1.3.3 and 1.3.4 and the successive approximations

$$x_0(t) = x_0, \qquad x_n(t) = x_0 + \int_{t_0}^t f[s, x_{n-1}(s)] \, ds, \qquad |t - t_0| \leqslant \alpha.$$

Justify passing to the limit under the integral sign.]

1.4. Abstract Lebesgue Integrals

Here we shall outline the Bochner and Pettis integrals which are defined relative to the strong and weak topology, respectively, on a Banach space X. These integrals are of the Lebesgue type. Let us begin with some notions. Let (Ω, S, m) be a measure space. The function $x : \Omega \to X$ is said to be

(i)   *countably valued* in Ω if it assumes at most a countable set of values in X, assuming each value different from zero on a measurable set;

(ii)   *weakly measurable* in Ω if the scalar function $\phi[x(\sigma)]$ is measurable for every $\phi \in X^*$;

(iii) *strongly measurable* in Ω if there exists a sequence $\{x_n(\sigma)\}_{n=1}^{\infty}$ of countably-valued functions (strongly) converging almost everywhere in Ω to $x(\sigma)$.

One can prove that in a separable space X the concepts of weak and strong measurability coincide.

DEFINITION 1.4.1.   A function $x: \Omega \to X$ is said to be *Pettis integrable* in Ω if for every $F \in S$ there exists a vector $x_F \in X$ such that for every $\phi \in X^*$

$$\phi(x_F) = \int_F \phi\,[x(\sigma)]\,dm \qquad (1.4.1)$$

where the integral in (1.4.1) is supposed to exist in the Lebesgue sense. By definition

$$(\text{P}) \int x(\sigma)\,dm = x_F.$$

It is not difficult to prove that in a reflexive space X the function $x: \Omega \to X$ is Pettis integrable if and only if $\phi\,[x(\sigma)]$ is Lebesgue integrable in Ω for every $\phi \in X^*$.

DEFINITION 1.4.2.   A countably valued function $x: \Omega \to X$ is called *Bochner integrable* in Ω if $\|x(\sigma)\|$ is Lebesgue integrable in Ω. By definition, for every F

$$(\text{B}) \int_F x(\sigma)\,dm = \sum_{i=1}^{\infty} x_i\,m(F_i \cap F) \qquad (1.4.2)$$

where $x_i = x(\sigma)$ for $\sigma \in F_i$, $i = 1, 2, \dots$.

The Bochner integral for countably valued functions is well defined since

$$\int_F \|x(\sigma)\|\,dm = \sum_{i=1}^{\infty} \|x_i\|\,m(F_i \cap F),$$

and so the series in (1.4.2) converges absolutely. It follows that

$$\left\| (\text{B}) \int_F x(\sigma)\,dm \right\| \leqslant \sum_{i=1}^{\infty} \|x_i\|\,m(F_i \cap F)$$

$$= (\text{L}) \int_F \|x(\sigma)\|\,dm.$$

Moreover

$$\phi\left[\int_F x(\sigma)\, dm\right] = \phi\left[\sum_{i=1}^{\infty} x_i\, m(F_i \cap F)\right]$$
$$= \sum_{i=1}^{\infty} \phi(x_i)\, m(F_i \cap F)$$
$$= \int_F \phi[x(\sigma)]\, dm,$$

that is, the Pettis and Bochner integrals for countably valued functions coincide.

DEFINITION 1.4.3.   A function $x: \Omega \to X$ is called *Bochner integrable* in Ω if there exists a sequence of countably valued, Bochner-integrable functions $x_n(\sigma)$ converging almost everywhere to $x(\sigma)$ and such that

$$\lim_{n\to\infty} \int_{\Omega} \|x_n(\sigma) - x(\sigma)\|\, dm = 0. \tag{1.4.3}$$

By definition

$$(\mathrm{B})\int_F x(\sigma)\, dm = \lim_{n\to\infty} (\mathrm{B})\int x_n(\sigma)\, dm. \tag{1.4.4}$$

We can establish that (1.4.3) is meaningful and that the Bochner integral is independent of the particular sequence $\{x_n(\sigma)\}$. The following theorem gives a necessary and sufficient condition for the existence of the Bochner integral of the function x.

THEOREM 1.4.1.   The function $x: \Omega \to X$ is Bochner integrable if and only if $x(\sigma)$ is strongly measurable and $\|x(\sigma)\|$ is Lebesgue integrable in Ω.

1.5. The Abstract Stieltjes Integral

Here we shall outline the abstract Stieltjes integral of a function $x: [a, b] \to X$ with respect to a function $y: [a, b] \to Y$. Let X, Y and Z be three Banach spaces. A bilinear operator $P: X \times Y \to Z$ whose norm is less than or equal to 1, that is, $\|P(x, y)\| \leqslant \|x\|\, \|y\|$, is called a *product operator*. We shall agree to write $P(x, y) = xy$. Let $x: [a, b] \to X$ and $y: [a, b] \to Y$ be two bounded functions such that the product $x(t)y(t) \in Z$, for each $t \in [a, b]$ is linear in both x and y and $\|x(t)y(t)\| \leqslant \|x(t)\|\, \|y(t)\|$ (for example, $x(t) = A(t)$ is an operator with domain $D[A(t)] \supset Y$, or one of the functions x, y is a scalar function). We denote the partition $(a = t_0 < t_1 < \cdots < t_n = b)$ together with the points τ_i ($t_i \leqslant \tau_i \leqslant t_{i+1}$, $i = 0, 1, \ldots, n-1$) by π and set $|\pi| = \max_i |t_{i+1} - t_i|$. We form the Stieltjes sum

$$S_\pi = \sum_{i=0}^{n-1} x(\tau_i)[y(t_{i+1}) - y(t_i)].$$

If the $\lim S_\pi$ exists as $|\pi| \to 0$ and defines an element I in Z independent of π, then I is called the *Stieltjes integral* of the function $x(t)$ by the function $y(t)$, and is denoted by

$$\int_a^b x(t)\, dy(t).$$

Notice that the Riemann integral of $x(t)$ which we defined in Section 1.3 is a special case of the Stieltjes integral, for the choice $y(t) \equiv t$.

We first need to introduce the concept of total variation for abstract functions. Consider the function $y: [a, b] \to X$ and the partition $\pi: a = t_0 < t_1 < \cdots < t_n = b$. Form the sum

$$V = \sum_{i=0}^{n-1} \|y(t_{i+1}) - y(t_i)\|.$$

The least upper bound of the set of all possible sums V is called the (*strong*) *total variation* of the function $y(t)$ on the interval $[a, b]$ and is denoted by $V_a^b(y)$. If $V_a^b(y) < \infty$, then $y(t)$ is called an abstract function of *bounded variation* on $[a, b]$. From Theorem 1.2.1, it follows that, if $y(t)$ has bounded derivative on $[a, b]$, then it is of bounded variation on $[a, b]$.

THEOREM 1.5.1. *If* $x \in C[[a, b], X]$ *and* $y: [a, b] \to Y$ *is of bounded variation on* $[a, b]$, *then the Stieltjes integral* $\int_a^b x(t)\, dy(t)$ *exists.*

The proof of this theorem makes use of the uniform continuity of $x(t)$ on $[a, b]$, the completeness of the space Z and the hypothesis that $\|x(t)y(t)\| \leqslant \|x(t)\| \|y(t)\|$. The details are left to the reader.

Most of the properties of the classical Stieltjes integral are also valid for the abstract Stieltjes integral and can be varified directly from the definition.

PROBLEM 1.5.1. *If* $x \in C[[a, b], X]$ *and* $y: [a, b] \to Y$ *is of bounded variation on* $[a, b]$, *then*

$$\left\| \int_a^b x(t)\, dy(t) \right\| \leqslant \int_a^b \|x(t)\|\, dV_a^t[y(t)]$$

$$\leqslant \max_{[a,b]} \|x(t)\|\, V_a^b[y(t)].$$

PROBLEM 1.5.2.   If $\{x_n(t)\}$ is a sequence of continuous abstract functions which converges uniformly to the necessarily continuous function $x(t)$ on the interval $[a, b]$ and if $y(t)$ is of bounded variation on $[a, b]$, then

$$\lim_{n \to \infty} \int_a^b x_n(t) \, dy(t) = \int_a^b x(t) \, dy(t).$$

1.6. Gateaux and Fréchet Differentials

Let X and Y be real Banach spaces and f be a mapping from an open set S of X into Y.

DEFINITION 1.6.1.   If for a fixed point $x \in S$ and every point $h \in X$ the

$$\lim_{t \to 0} [f(x + th) - f(x)]/t = \delta f(x, h)$$

exists, in the topology of Y, then the operator $\delta f(x, h)$ is called the *Gateaux differential* of the function f at the point x in the direction h.

The Gateaux differential generalizes the concept of directional derivative familiar in finite-dimensional spaces. (Actually, the existence of Gateaux differential does not require a norm on X.) For a fixed $x \in S$ and h regarded as a variable, the Gateaux differential defines a transformation from X into Y. In the special case that f is linear $\delta f(x, h) = f(h)$. The example $f: R^2 \to R$ defined by

$$f(x, y) = x^3/(x^2 + y^2), \qquad (x, y) \neq (0, 0) \qquad \text{and} \qquad f(0, 0) = 0$$

shows that $\delta f(x, h)$ is not always linear in h. However, $\delta f(x, h)$ is always homogeneous in h. In fact, $\delta f(x, 0) = 0$ and for $\lambda \neq 0$, setting $\tilde{t} = \lambda t$, we have

$$\delta f(x, \lambda h) = \lim_{t \to 0} [f(x + t\lambda h) - f(x)]/t$$

$$= \lambda \lim_{\tilde{t} \to 0} [f(x + \tilde{t} h) - f(x)]/\tilde{t}$$

$$= \lambda \delta f(x, h).$$

EXAMPLE 1.6.1.   If $f: R^n \to R$ where $f(x) = f(x_1, x_2, \ldots, x_n)$ has continuous partial derivatives with respect to each variable x_i, then

$$\delta f(x, h) = d/dt \, [f(x + th)]_{t=0} = \sum_{i=1}^{n} [\partial f(x)/\partial x_i] h_i.$$

A constant function has Gateaux differential equal to zero. The converse is also true as we will see in Corollary 1.6.1. In a Hilbert space H, let $f: H \to R_+$ be given by $f(x) = \|x\|$. Then

$$\delta f(x, h) = (x, h)/\|x\| \qquad \text{for} \quad x \neq 0.$$

Indeed

$$(\|x + th\| - \|x\|)/t = (\|x + th\|^2 - \|x\|^2)/t(\|x + th\| + \|x\|)$$

$$= (2(x, h) + t\|h\|^2)/(\|x + th\| + \|x\|)$$

$$\to (x, h)/\|x\| \qquad \text{as} \quad t \to 0.$$

If $f: H \to R_+$ is given by $f(x) = (x, x)$, then clearly $\delta f(x, h) = 2(x, h)$. Finally, if $f: C[0, 1] \to R$ defined by

$$f(x) = \int_0^1 g[s, x(s)] \, ds$$

where g_x exists and is continuous with respect to s and x, then

$$\delta f(x, h) = (d/dt) \left(\int_0^1 g[s, x(s) + th(s)] \, ds \right)_{t=0}$$

$$= \int_0^1 g_x[s, x(s)] h(s) \, ds.$$

It is well known that the mean value theorem is not true for vector-valued functions. However, it is true for functionals $\phi: S \to R$ where S is an open set in X as the following theorem proves.

THEOREM 1.6.1. Let the Gateaux differential $\delta\phi(x, h)$ of a functional $\phi: S \to R$ exist for each point of a convex set $V \subset S$ (and any direction in X). Then for any pair of points $x, x + h \in V$ there exists a number $\tau \in (0, 1)$ such that

$$\phi(x + h) - \phi(x) = \delta\phi(x + \tau h, h). \tag{1.6.1}$$

Proof: Set $F(t) = \phi(x + th)$. Then $F: [0, 1] \to R$ and

$$F'(t) = \lim_{\Delta t \to 0} [\phi(x + th + h\Delta t) - \phi(x + th)]/\Delta t$$

$$= \delta\phi(x + th, h), \qquad 0 < t < 1, \tag{1.6.2}$$

which exists since $x + th \in V$ for all $t \in [0, 1]$. The classical mean value theorem applied to F gives $F(1) - F(0) = F'(\tau)$ for some $\tau \in (0, 1)$ and (1.6.1) follows.

For functions $f: S \to Y$ which have Gateaux differentials the following form of the mean value theorem is valid.

THEOREM 1.6.2.   Let the Gateaux differential $\delta f(x, h)$ of a function $f: S \to Y$ exist at each point of a convex set $V \subset S$ (and any direction in X). Then for any pair of points x, $x + h \in V$ there exists a number $\tau \in (0, 1)$ such that

$$\|f(x+h) - f(x)\| \leqslant \|\delta f(x+\tau h, h)\|. \tag{1.6.3}$$

Proof: Let $\phi \in X^*$. Define the functional $\tilde{\phi}(x) = \phi[f(x)]$. Then the Gateaux differential of $\tilde{\phi}$ exists at each point in V (and any direction in X). In fact, since $\phi \in X^*$

$$[\tilde{\phi}(x+th) - \tilde{\phi}(x)]/t = \phi[f(x+th) - f(x)]/t$$
$$\to \phi[\delta f(x, h)] \qquad \text{as} \quad t \to 0.$$

In view of Theorem 1.6.1 there exists a point $\tau \in (0, 1)$ such that

$$\tilde{\phi}(x+h) - \tilde{\phi}(x) = \delta\tilde{\phi}(x+\tau h, h)$$
$$= \phi[\delta f(x+\tau h, h)]. \tag{1.6.4}$$

If the vector $f(x+h) - f(x) = 0$, then (1.6.3) is clearly valid. If $f(x+h) - f(x) \neq 0$ for the pair of points (x, h), then we can choose the functional ϕ such that

$$\phi[f(x+h) - f(x)] = \|f(x+h) - f(x)\| \qquad \text{and} \qquad \|\phi\| = 1.$$

(Such a choice is possible by means of Appendix III.) Then, using (1.6.4)

$$\|f(x+h) - f(x)\| = |\phi[f(x+h) - f(x)]|$$
$$= |\phi[\delta f(x+\tau h, h)]|$$
$$\leqslant \|\delta f(x+\tau h, h)\| \|\phi\|$$
$$\leqslant \|\delta f(x+\tau h, h)\|$$

and the proof is complete.

COROLLARY 1.6.1.   If $\delta f(x, h) = 0$ for all points x in a convex set $V \subset X$ then $f(x) \equiv \text{const}$ on V.

As we have seen $\delta f(x, h)$ is not always linear in h. A set of sufficient conditions, which guarantee the linearity in h of the Gateaux differential $\delta f(x, h)$, is given in the following:

PROBLEM 1.6.1. Assume that

(i) $\delta f(x, h)$ exists in a neighborhood of x_0 and is continuous in x at the point x_0;
(ii) $\delta f(x_0, h)$ is continuous in h at the point $h = 0$. Then, $\delta f(x, h)$ is linear in h.

[Hint: Use the homogeneity in h and (1.6.4) with $\|\phi\| = 1$.]

DEFINITION 1.6.2. If for a fixed point $x \in X$ the Gateaux differential $\delta f(x, \cdot)$ is a bounded linear operator mapping X into Y, we write $\delta f(x, h) \equiv f'(x)h$ and $f'(x)$ is called the *Gateaux derivative* of f at x. In the special case $Y = R, f'(x)$ is called the *gradient* of the functional f at the point x.

In a real Hilbert space H the gradient at the point $x \neq 0$ of the functional $f: H \to R_+$ defined by $f(x) = \|x\|$ is $(x/\|x\|, \cdot)$ or simply $x/\|x\|$. If $f: H \to R_+$ is given by $f(x) = \|x\|^2$, then the gradient at any point x is $2(x, \cdot)$ or simply $2x$.

When the function f has a Gateaux derivative at the point x, we say that f is G-differentiable at x. The function f is called G-differentiable in a set $A \subset X$ if it is G-differentiable at every point of A.

From Definition 1.6.2. the Gateaux derivative $f'(x)$ of the function $f: S \to Y$, if it exists, is an element of $B(X, Y)$ where $B(X, Y)$ is the space of bounded linear operators from X into Y.

A more satisfactory differential concept which requires a norm on X (the Gateaux differential does not) is the following:

DEFINITION 1.6.3. Let $f: S \to Y$ be a function from an open set S of the Banach space X into the Banach space Y. If at a point $x \in S$

$$f(x+h) - f(x) = L(x, h) + w(x, h), \qquad h \in X$$

where $L(x, \cdot): X \to Y$ is a linear operator and

$$\lim_{\|h\| \to 0} \|w(x, h)\|/\|h\| = 0, \tag{1.6.5}$$

then $L(x, h)$ is called the *Fréchet differential* of the function f at the point x with *increment* h and $w(x, h)$ is called the *remainder* of the differential. The operator $L(x, \cdot): X \to Y$ is called the *Fréchet derivative* of f at x, and is denoted by $f'(x)$.

LEMMA 1.6.1. The Fréchet differential of a function f at the point x, if it exists, is unique.

Proof: Let $L(x,h)$ and $\tilde{L}(x,h)$ be both Fréchet differentials of the function f at the point x with remainders $w(x,h)$ and $\tilde{w}(x,h)$ respectively. Then

$$L(x,h) + w(x,h) = \tilde{L}(x,h) + \tilde{w}(x,h).$$

Therefore

$$\|L(x,h) - \tilde{L}(x,h)\|/\|h\| = \|\tilde{w}(x,h) - w(x,h)\|/\|h\|$$

$$\leqslant \|\tilde{w}(x,h)\|/\|h\| + \|w(x,h)\|/\|h\| \to 0$$

as $\|h\| \to 0$.

Set $Th = L(x,h) - \tilde{L}(x,h)$. Then T is a linear operator from X into Y such that

$$\lim_{\|h\|\to 0} \|Th\|/\|h\| = 0.$$

Hence, for every $\varepsilon > 0$ there exists a $\delta = \delta(\varepsilon) > 0$ such that $\|h\| < \delta$ implies $\|Th\|/\|h\| < \varepsilon$. Observe that the vector $v = \delta x/2\|x\|$ satisfies $\|v\| < \delta$, therefore $\|Tv\|/\|v\| < \varepsilon$ and using the definition of v we obtain $\|Tx\| < \varepsilon\|x\|$ for any $x \in X$ and any $\varepsilon > 0$. From this it follows that $T \equiv 0$. Otherwise, there exists an $x_0 \in X$ such that $Tx_0 \neq 0$. Then $\|Tx_0\| \leqslant \varepsilon\|x_0\| \to 0$ as $\varepsilon \to 0$, contradicting the hypothesis. Hence $L(x,h) = \tilde{L}\tilde{L}(x,h)$ for all $h \in X$ and the proof is complete.

When the function f has Fréchet derivative at the point x, we say that f is F-differentiable at x. The function f is called F-differentiable in a set $A \subset X$ if it is F-differentiable at every point of A.

LEMMA 1.6.2.   If the continuous function $f: S \to Y$ has Fréchet differential $L(x_0, h)$ at the point $x_0 \in S$, then $L(x_0, \cdot): X \to Y$ is a bounded linear operator, that is, the Fréchet derivative $f'(x_0)$ of f at x_0 is an element of $B(X, Y)$.

Proof: In view of the continuity of f and (1.6.5), it follows that for a given $\varepsilon > 0$ there exists a $\delta \in (0, 1)$ such that $\|h\| < \delta$ implies

$$\|f(x_0 + h) - f(x_0)\| < \varepsilon/2$$

and

$$\|f(x_0 + h) - f(x_0) - L(x_0, h)\| < (\varepsilon/2)\|h\| \leqslant \varepsilon/2.$$

Hence, for $\|h\| < \delta$

$$\|L(x_0, h)\| < \varepsilon/2 + \|f(x_0 + h) - f(x_0)\| < \varepsilon, \tag{1.6.6}$$

which proves that $L(x_0, \cdot)$ is bounded by $2\varepsilon/\delta$. In fact, for an arbitrary $\tilde{h} \in X$ set $h = \delta\tilde{h}/2\|\tilde{h}\|$ so that $\|h\| < \delta$. By (1.6.6) and the linearity of $L(x_0, h)$ in h, it follows that

$$\varepsilon > \|L(x_0, h)\| = \|L(x_0, \delta\tilde{h}/2\|\tilde{h}\|)\|$$
$$= (\delta/2)\|L(x_0, \tilde{h})\|/\|\tilde{h}\|,$$

which proves our assertion.

LEMMA 1.6.3.   If the Fréchet differential of the function f at x_0 exists then the Gateaux differential exists at x_0 and they are equal.

Proof: From the definition of Fréchet differential we have

$$f(x_0 + ht) - f(x_0) = L(x_0, ht) + w(x_0, ht).$$

Thus, by the linearity of $L(x, \cdot)$ and the property of the remainder, we obtain

$$\delta f(x_0, h) = \lim_{t \to 0} [f(x_0 + ht) - f(x_0)]/t$$
$$= L(x_0, h)$$

and the proof is complete.

The converse of Lemma 1.6.3. is not always true. However, the following is true:

PROBLEM 1.6.2.   If the Gateaux derivative $f'(x)$ exists in a neighborhood $N(x_0)$ of the point x_0 and is continuous at x_0, then the Fréchet derivative exists and is equal to $f'(x)$. That is, a continuous Gateaux derivative is a Fréchet derivative.

[Hint: Use the mean value theorem for Gateaux differentials.]

One can easily verify that in Examples 1.6.1 all the Gateaux differentials are also Fréchet differentials. We shall prove this only for the first example $f: R^n \to R$ where $f(x) = f(x_1, x_2, \ldots, x_n)$ has continuous partial derivatives with respect to each variable x_i. The claim is that for this function

$$L(x, h) = \sum_{i=1}^{n} [\partial f(x)/\partial x_i] h_i.$$

Clearly, $L(x, h)$ is linear in h. The continuity of the partial derivatives implies that given $\varepsilon > 0$, there exists a neighborhood $N(x, \delta) = \{y \in R^n:$

$\|y - x\| < \delta\}$ such that $y \in N(x, \delta)$ and for any $i = 1, 2, ..., n$

$$|\partial f(x)/\partial x_i - \partial f(y)/\partial x_i]| < \varepsilon/n. \tag{1.6.7}$$

Define the unit vectors e_i in the usual way, $e_i = (0, ..., 0, 1, 0, ..., 0)$ and for $h = (h_1, h_2, ..., h_n)$ define

$$g_0 = (0, ..., 0) \quad \text{and} \quad g_k = \sum_{i=1}^{k} h_i e_i, \quad k = 1, 2, ..., n.$$

Notice that

$$\|g_k\| \leqslant \|h\| \quad \text{for all} \quad k = 1, 2, ..., n.$$

Then

$$\left| f(x+h) - f(x) - \sum_{i=1}^{n} [\partial f(x)/\partial x_i] h_i \right|$$

$$= \left| \sum_{i=1}^{n} [f(x+g_i) - f(x+g_{i-1}) - [\partial f(x)/\partial x_i] h_i] \right|$$

$$\leqslant \sum_{i=1}^{n} |f(x+g_i) - f(x+g_{i-1}) - [\partial f(x)/\partial x_i] h_i|. \tag{1.6.8}$$

Employing the mean value theorem for functions of a single variable, there exists a constant τ such that

$$f(x+g_i) - f(x+g_{i-1}) = (\partial f/\partial x_i)(x+g_{i-1}+\tau e_i) h_i, \quad 0 \leqslant \tau \leqslant h_k, \tag{1.6.9}$$

and

$$x + g_{i-1} + \tau e_i \in N(x, \delta) \quad \text{for} \quad \|h\| < \delta.$$

From (1.6.9) and (1.6.7) we get

$$|f(x+g_i) - f(x+g_{i-1}) - [\partial f(x)/\partial x_i] h_i| < (\varepsilon/n) |h_i|.$$

This together with (1.6.9) yields for $\|h\| < \delta$

$$\left| f(x+h) - f(x) - \sum_{i=1}^{n} [\partial f(x)/\partial x] h_i \right| < \varepsilon \|h\|,$$

and (1.6.5) is valid. The claim is therefore established.

Much of the theory of ordinary derivatives can be generalized to Fréchet derivatives. For example, the implicit function theorem and Taylor series have very satisfactory extensions (see [15, 28]). The following properties can be established immediately from the definition:

(i)
$$[cf(x)]' = cf'(x)$$

where c is a real constant;

(ii)
$$[f_1(x) + f_2(x)]' = f_1'(x) + f_2'(x).$$

Next we establish the *chain rule* for Fréchet derivatives.

LEMMA 1.6.4. Let X, Y, Z be three Banach spaces; A an open neighborhood of $x_0 \in X, f \in C[A, Y], y_0 = f(x_0)$; B an open neighborhood of y_0 in Y and $g \in C[B, Z]$. Assume that f is F-differentiable at x_0 and g is F-differentiable at y_0. Then the function $h = g \circ f$ (which is defined and continuous in a neighborhood of x_0) is F-differentiable at x_0, and $h'(x_0) = g'(y_0) \circ f'(x_0)$ (where $g'(y_0) \circ f'(x_0)$ is understood as the product of the operators $g'(y_0): Y \to Z, f'(x_0): X \to Y$).

Proof: From the hypotheses, given $\varepsilon > 0, 0 < \varepsilon < 1$, there exists a $\delta > 0$ such that $\|h\| < \delta$ and $\|t\| \leqslant \delta$ imply

$$f(x_0 + h) - f(x_0) = f'(x_0)h + O_1(h)$$

and

$$g(y_0 + t) - g(y_0) = g'(y_0)t + O_2(t)$$

where $\|O_1(h)\| \leqslant \varepsilon \|h)$ and $\|O_2(t)\| \leqslant \varepsilon \|t\|$. In view of Lemma 1.6.2 the operators $f'(x_0)$ and $g'(y_0)$ are bounded and therefore there are constants M and N such that

$$\|f'(x_0)\| \leqslant M \quad \text{and} \quad \|g'(y_0)\| \leqslant N.$$

Also

$$\|f'(x_0)h + O_1(h)\| \leqslant (M+1)\|h\| \quad \text{for} \quad \|h\| < \delta.$$

Hence, for $\|h\| < \delta/(M+1)$, we have

$$\|O_2[f'(x_0)h + O_1(h)]\| \leqslant \varepsilon \|f'(x_0)h + O_1(h)\| \leqslant \varepsilon(M+1)\|h\|$$

and

$$\|g'(y_0)O_1(h)\| \leqslant N\varepsilon \|h\|.$$

On the strength of these inequalities one can write

$$h(x_0+h) = g[f(x_0+h)]$$
$$= g[f(x_0)+f'(x_0)h + O_1(h)]$$
$$= g[y_0 +f'(x_0)h + O_1(h)]$$
$$= g(y_0) + g'(y_0)[f'(x_0)h + O_1(h)] + O_2[f'(x_0)h + O_1(h)]$$
$$= g(y_0) + g'(y_0)f'(x_0)h + O_3(\|h\|) \qquad (1.6.10)$$

with

$$\|O_3(h)\| = \|g'(y_0)O_1(h) + O_2[f'(x_0)h + O_1(h)]\|$$
$$\leqslant N\varepsilon\|h\| + \varepsilon(M+1)\|h\|$$
$$= (M+N+1)\varepsilon\|h\|.$$

The identity (1.6.10) yields the desired result.

For F-differentiable functions Theorem 1.6.2. gives the following:

COROLLARY 1.6.2.   Let $f \in C[S, Y]$ and F-differentiable at each point of a convex set $V \subset S$. Then for any pair of points $x, x+h \in V$

$$\|f(x+h) - f(x)\| \leqslant \|h\| \sup_{0 < t < 1} \|f'(x+th)\|. \qquad (1.6.11)$$

1.7. Notes

Most of the results of this chapter are taken from Vainberg [75] and Hille and Phillips [28]. More details about Bochner integrals and abstract functions can be found in Hille and Phillips [28]. The proof of Lemma 1.6.4 is due to Dieudonné [16]. For other mean value theorems used in connection with differential equations the reader is referred to Aziz and Diaz [5].

Chapter 2

Semigroups of Operators

2.0. Introduction

The evolution of a physical system in time is described by an initial value problem of the form

$$du/dt = Au(t), \qquad t \geqslant 0 \quad \text{and} \quad u(0) = u_0 \qquad (2.0.1)$$

where $A: D(A) \to X$ is a linear operator with domain $D(A) \subset X$, X being a Banach space, $u: [0, \infty) \to X$ and $u_0 \in D(A)$. Here A does not depend on time. Physically this means that the underlying mechanism does not depend on time. We shall use the initial value problem (2.0.1) to motivate the theory of one-parameter semigroups of operators. Following the usage of Hadamard one calls the problem *well set* if

(i)   there is a unique solution to the problem for some given class of initial data;

(ii)   the solution varies continuously with the initial data.

These two requirements are reasonable to expect if (2.0.1) is to correspond to a well-set physical experiment. The existence and uniqueness of the solution is an affirmation of the *principle of scientific determinism* [28]; while the continuous dependence is an expression of the stability of the solution.

Suppose that (2.0.1) is well set and let $T(t)$ map the solution $u(s)$ at time s to the solution $u(t+s)$ at time $t+s$. Since A does not depend on time, the operator $T(t)$ does not depend on s. The stability requirement implies that $T(t)$ is a continuous operator on X. The solution $u(t+\tau)$ at time $t+\tau$ is given by $T(t+\tau)u_0$. At time τ the solution is $T(\tau)u_0$. Therefore taking this as initial data t units of time later, the solution becomes $u(t+\tau) = T(t)[T(\tau)u_0]$. From the uniqueness requirement and assuming that $D(A)$ is dense in X, we obtain the semigroup property

$$T(t+s) = T(t)\,T(s), \qquad t, s > 0.$$

Since the initial condition in (2.0.1) must be satisfied, we must have $\lim_{t\to 0_+} T(t)u_0 = u_0$. In other words, the operators $T(t)$ converge strongly to the identity operator as $t\to 0_+$.

The foregoing discussion shows how the initial value problem (2.0.1), when assumed to be well set, leads to the concept of one parameter semigroup $\{T(t)\}$, $t \geqslant 0$, of bounded linear operators on a Banach space X. When the operator A in (2.0.1) is a matrix in R^n, the solution of (2.0.1) is given by

$$u(t) \equiv T(t)u_0 = \exp(tA)u_0.$$

By analogy with the matrix case we could expect that a semigroup $\{T(t)\}$ is, in some sense, an exponential function even when A is an unbounded operator. A is called the infinitesimal generator of the semigroup $\{T(t)\}$ and in a sense, which will be made precise later, $T(t) = \exp(tA)$, $t \geqslant 0$.

This chapter is therefore devoted to the exposition of the most basic results of semigroup theory as developed by E. Hille, R. S. Phillips, and K. Yosida. Several examples and problems are given to illustrate the concepts and their applications to partial and functional differential equations. The abstract Cauchy problem (2.0.1) and its corresponding non-homogeneous problem is also carefully studied when A generates a strongly continuous semigroup, as well as an analytic semigroup.

2.1. Strongly Continuous Semigroups of Operators

DEFINITION 2.1.1.   A family $\{T(t)\}, 0 \leqslant t < \infty$, of bounded linear operators mapping the Banach space X into X is called a *strongly continuous semigroup* of operators if the following three conditions are satisfied:

(i) $\qquad\qquad T(t+s) = T(t)\,T(s), \qquad t,s \geqslant 0;$

(ii) $\qquad T(0) = I \qquad$ (I is the identity operator in X);

(iii)   for each $x \in X$, $T(t)x$ is (strongly) continuous in t on $[0, \infty)$, that is,

$$\|T(t+\Delta t)\,x - T(t)\,x\| \to 0 \qquad \text{as} \quad \Delta t \to 0, \qquad t, t+\Delta t \geqslant 0.$$

If in addition to the conditions (i), (ii) and (iii) the map $t \to T(t)$ is continuous in the uniform operator topology, that is, $\|T(t+\Delta t) - T(t)\| \to 0$ for $t, t+\Delta t \geqslant 0$, then the family $\{T(t)\}, t \geqslant 0$, is called a *uniformly continuous semigroup* in X. If the strongly continuous semigroup $\{T(t)\}, t \geqslant 0$, satisfies the property $\|T(t)\| \leqslant 1$ for $t \geqslant 0$, then it is called a *contraction semigroup* in X.

We observe that the operators $T(t)$ and $T(s)$ commute as a consequence of (i). Property (ii) does not follow from (i), and $T(t)x$ is also (strongly) continuous in x for each $t \geqslant 0$.

EXAMPLE 2.1.1.   Let A be a bounded linear operator in a Banach space X, that is, any linear operator in a finite dimensional unitary space. Then the series $\sum_{n=0}^{\infty} (A^n/n!)\, t^n$ converges in the uniform operatory topology, that is, in the norm of $B(X)$, for any real number t. In fact, set

$$S_n = \sum_{k=0}^{n} (A^k/k!)\, t^k$$

and observe that for $m < n$

$$\|S_n - S_m\| = \left\| \sum_{k=m+1}^{n} (A^k/k!)\, t^k \right\| \leqslant \sum_{k=m+1}^{n} (\|A\|^k/k!)\, |t|^k \to 0$$

$$\text{as} \quad m, n \to \infty.$$

So $\{S_n\}$ is a Cauchy sequence in the Banach space $B(X)$ and consequently it converges to an operator in $B(X)$ which we denote by $\exp(tA)$. Now we can easily verify that the family of bounded operators $\{\exp(tA)\}, t \geqslant 0$, is a uniformly continuous semigroup in X. In addition we can show that

$$(d/dt)\exp(tA) = A\exp(tA).$$

PROBLEM 2.1.1. Let A be a bounded operator in the Banach space X. Show that the Cauchy problem

$$dx/dt = Ax, \qquad t \geqslant 0 \quad \text{and} \quad x(0) = x_0 \in X$$

has the unique solution $x(t) = \exp(tA) x_0$. (For uniqueness see Theorem 2.1.2.)

It is possible to show [18] that if $\{T(t)\}$, $t \geqslant 0$, is a uniformly continuous semigroup of operators, then there exists a bounded linear operator A such that $T(t) = \exp(tA)$, $t \geqslant 0$. The operator A is given by the formula

$$A = \lim_{h \to 0_+} [T(h) - I]/h$$

where the limit is taken in the uniform operator topology.

EXAMPLE 2.1.2. In the Banach space $X = C[0, 1]$ of continuous functions with sup-norm, define the family of operators $\{T(t)\}$, $t \geqslant 0$ by the formula

$$T(t) x(\xi) = x[\xi/(1 + t\xi)], \qquad x \in X, \quad \xi \in [0, 1].$$

Then $\{T(t)\}$, $t \geqslant 0$, is a strongly continuous semigroup of operators in X. Indeed

(i) $$T(t) T(s) x(\xi) = T(t) x[\xi/(1 + s\xi)]$$

$$= x\left(\frac{\xi/(1 + s\xi)}{1 + [t\xi/(1 + s\xi)]}\right)$$

$$= x(\xi/[1 + (t+s)\xi])$$

$$= T(t+s) x(\xi);$$

(ii) $$T(0) x(\xi) = x(\xi);$$

(iii) $$\|T(t+\Delta t) x - T(t) x\| = \sup_{0 \leqslant \xi \leqslant 1} |x(\xi/[1 + (t+\Delta t)\xi])$$

$$- x[\xi/(1 + t\xi)]| \to 0 \qquad \text{as} \quad \Delta t \to 0.$$

EXAMPLE 2.1.3. In the Banach space $X = C[0, \infty)$ of continuous, bounded functions on $[0, \infty)$ with sup-norm, define the family of translation operators $\{T(t)\}$, $t \geqslant 0$, by the formula

$$T(t) x(\xi) = x(t+\xi), \qquad x \in X, \quad \xi \geqslant 0.$$

Then $\{T(t)\}$, $t \geqslant 0$, is a strongly continuous, contraction semigroup of operators in X.

EXAMPLE 2.1.4.   Let A on $D(A)$ be a self-adjoint operator in the Hilbert space H. Assume that $(Ax, x) \leqslant 0$ for $x \in D(A)$, that is, A is negative. Let $\{E_\lambda\}$, $-\infty < \lambda < \infty$, be the resolution of the identity for the operator A. Then $E_\lambda = I$ for $\lambda > 0$ and $Ax = \int_{-\infty}^{0} \lambda \, dE_\lambda x$ for $x \in D(A)$.

Define the family of operators $\{T(t)\}$, $t \geqslant 0$, by

$$T(t)x = \int_{-\infty}^{0} \exp(\lambda t) \, dE_\lambda x, \qquad x \in H.$$

This family of operators is a strongly continuous semigroup in H.

We shall now prove that the norm $\|T(t)\|$ of the operators in a strongly continuous semigroup grows slower than an exponential. For this we need the concept of subadditive function and a lemma on subadditive functions. A function $\omega: [0, \infty) \to R$ is called *subadditive* if

$$\omega(t_1 + t_2) \leqslant \omega(t_1) + \omega(t_2), \qquad t_1, t_2 \geqslant 0.$$

For example, the function $\omega(t) = \log \|T(t)\|$ for $t \geqslant 0$ is a subadditive function.

LEMMA 2.1.1.   Let $\omega: [0, \infty) \to R$ be subadditive and bounded above on each finite subinterval. Then the number $\omega_0 = \inf_{t>0} \omega(t)/t$ is finite or $-\infty$ and $\omega_0 = \lim_{t \to \infty} \omega(t)/t$.

Proof: Let $\omega_0 = \inf_{t>0} \omega(t)/t$. Since $\omega(t)$ is bounded above, ω_0 is finite or $-\infty$. Suppose that ω_0 is finite. Given any $\delta > \omega_0$ there exists a t_0 such that $\omega(t_0)/t_0 < \delta$. For any $t \geqslant 0$, we can write

$$t = n(t)t_0 + r$$

where $n(t)$ is an integer and $0 \leqslant r < t_0$. Then

$$\omega(t)/t = \omega[n(t)t_0 + r]/t$$
$$\leqslant [n(t)\omega(t_0) + \omega(r)]/t$$
$$= \frac{\omega(t_0)}{[t_0 + r/n(t)]} + \frac{\omega(r)}{t}.$$

Thus

$$\limsup_{t \to \infty} \omega(t)/t \leqslant \omega(t_0)/t_0 < \delta. \qquad (2.1.1)$$

From the definition of ω_0 it follows that

$$\omega_0 \leqslant \liminf_{t \to \infty} \omega(t)/t. \qquad (2.1.2)$$

The relations (2.1.1) and (2.1.2) yield the desired result.

THEOREM 2.1.1.   The limit

$$\omega_0 = \lim_{t \to \infty} (\log \|T(t)\|)/t$$

exists. For each $\delta > \omega_0$ there exists a constant M_δ such that

$$\|T(t)\| \leqslant M_\delta \exp(\delta t), \qquad t \geqslant 0.$$

The number ω_0 is called the *type* of the semigroup.

Proof: Define the subadditive function

$$\omega(t) = \log \|T(t)\|, \qquad t \geqslant 0.$$

We first prove that the function $\|T(t)\|$ is bounded for t in a finite interval $[0, t_0]$. If not, there exists a sequence $t_n \to t^* \in [0, t_0]$ for $t_n \in [0, t_0]$ such that $\|T(t_n)\| \to \infty$ while $\|T(t^*)\|$ is a finite number. For every $x \in X$ we have $T(t_n)x \to T(t^*)x$ as $n \to \infty$. Hence $\sup_n \|T(t_n)x\| < \infty$ for each $x \in X$. By the uniform boundedness principle (see Appendix VI) we conclude that $\sup_n \|T(t_n)\| < \infty$, which is a contradiction. In view of Lemma 2.1.1, we have that $\omega_0 = \lim_{t \to \infty} (\log \|T(t)\|)/t$ exists and is a finite number or $-\infty$. For any $\delta > \omega_0$ there is a t_0 such that

$$(\log \|T(t)\|)/t < \delta, \qquad t \geqslant t_0.$$

Hence

$$\|T(t)\| \leqslant \exp(\delta t), \qquad t \geqslant t_0.$$

In addition we know that $\|T(t)\|$ is bounded on $[0, t_0]$ and the result follows.

2.2. The Infinitesimal Generator

Let $\{T(t)\}$, $t \geqslant 0$, be a strongly continuous semigroup of operators in the Banach space X. For $h > 0$ we define the linear operator A_h by the formula

$$A_h x = [T(h)x - x]/h, \qquad x \in X.$$

Let $D(A)$ be the set of all $x \in X$ for which the $\lim_{h \to 0_+} A_h x$ exists. Define the operator A on $D(A)$ by the relation

$$Ax = \lim_{h \to 0_+} A_h x, \qquad x \in D(A).$$

DEFINITION 2.2.1. The operator A with domain $D(A)$ is called the *infinitesimal generator* of the semigroup $\{T(t)\}$, $t \geq 0$. Given an operator A on $D(A)$, we say that it *generates* a strongly continuous semigroup $\{T(t)\}$, $t \geq 0$ if A coincides with the infinitesimal generator of $\{T(t)\}$, $t \geq 0$.

PROBLEM 2.2.1. Find the infinitesimal generators of the semigroups in Examples 2.1.1–2.1.4.

[Answers: A; $-\xi^2 (d/d\xi)$; $(d/d\xi)$; A.]

The following properties of the infinitesimal generator are very useful.

THEOREM 2.2.1. Let $\{T(t)\}$, $t \geq 0$ be a strongly continuous semigroup of operators in the Banach space X and A its infinitesimal generator with domain $D(A)$. Then

(a) $D(A)$ is a linear manifold in X and A on $D(A)$ is a linear operator;
(b)   if $x \in D(A)$, then $T(t)x \in D(A)$, $0 \leq t < \infty$, is (strongly) differentiable in t and

$$(d/dt) T(t) x = AT(t) x$$
$$= T(t) Ax, \qquad t \geq 0;$$

(c)   if $x \in D(A)$, then

$$T(t)x - T(s)x = \int_s^t T(u) Ax \, du, \qquad t, s \geq 0;$$

(d)   if $f(t)$ is a continuous real-valued function for $t \geq 0$, then

$$\lim_{h \to 0} h^{-1} \int_t^{t+h} f(u) T(u) x \, du = f(t) T(t) x, \qquad x \in X, \quad t \geq 0;$$

(e) $\displaystyle\int_0^t T(s) x \, ds \in D(A)$ and $\displaystyle T(t)x = x + A \int_0^t T(s) x \, ds,$

$$x \in X;$$

(f)   the linear manifold $D(A)$ is dense in X, and A on $D(A)$ is a closed operator.

Proof: (a)   Follows directly from the definition.
(b)   Since $T(t)$ and $T(h)$ commute, we have

$$A_h T(t) x = T(t) A_h x$$
$$\to T(t) Ax \qquad \text{as} \quad h \to 0_+.$$

Hence

$$T(t)x \in D(A) \quad \text{and} \quad AT(t)x = T(t)Ax.$$

Next we will show that $(d/dt)[T(t)x] = T(t)Ax$ by showing that this is true both for the left- and right-hand derivatives. Indeed

$$\lim_{h \to 0_+} ([T(t+h)x - T(t)x]/h - T(t)Ax) = \lim_{h \to 0_+} [T(t)A_h x - T(t)Ax] = 0$$

and

$$\lim_{h \to 0_-} ([T(t+h)x - T(t)x]/h - T(t)Ax)$$

$$= \lim_{h \to 0_-} [T(t+h)\{([T(-h)x - x]/-h) - Ax\} + [T(t+h) - T(t)]Ax] = 0.$$

(c)   The abstract function $y(t) = T(t)x$ is differentiable by (b) and its derivative $T(t)Ax$ is continuous in t. The conclusion follows from Theorem 1.3.4.

(d)   The abstract function $y(t) = f(t)T(t)x$ is continuous in t. Set

$$F(\xi) = \int_t^{t+\xi} y(s) \, ds.$$

Then

$$F'(0) = \lim_{h \to 0} h^{-1} \int_t^{t+h} y(s) \, ds.$$

On the other hand, by Theorem 1.3.3 we get

$$F'(\xi) = y(t+\xi).$$

Hence

$$F'(0) = f(t)T(t)x$$

and the result follows by equating the values of $F'(0)$.

(e)   Let $x \in X$ and $t, h > 0$. Then, as $h \to 0$

$$A_h \int_0^t T(s)x \, ds = h^{-1} \int_0^t [T(h+s)x - T(s)x] \, ds$$

$$= h^{-1} \int_h^{t+h} T(s)x \, ds - h^{-1} \int_0^t T(s)x \, ds$$

$$= h^{-1} \int_t^{t+h} T(s)x \, ds - h^{-1} \int_0^h T(s)x \, ds$$

$$\to T(t)x - T(0)x$$

$$= T(t)x - x \quad \text{(from (d))}.$$

This proves that

$$\int_0^t T(s)x\,ds \in D(A) \qquad \text{and} \qquad A\int_0^t T(s)x\,ds = T(t)x - x.$$

(f)   From (d) $x = \lim_{t\to 0} t^{-1}\int_0^t T(s)x\,ds$ for every $x \in X$ and from (e) $\int_0^t T(s)x\,ds \in D(A)$ for every $x \in X$. These two facts imply that $D(A)$ is dense in X. Finally we show that A on $D(A)$ is a closed operator. Let $x_n \in D(A)$ with $n = 1, 2, \ldots$, $\lim_{n\to\infty} x_n = x$ and $\lim_{n\to\infty} Ax_n = y$. We must prove that $x \in D(A)$ and that $y = Ax$. By (c) and the fact that $T(s)Ax_n \to T(s)y$ uniformly we get

$$T(t)x - x = \lim_{n\to\infty} [T(t)x_n - x_n]$$

$$= \lim_{n\to\infty} \int_0^t T(s)Ax_n\,ds$$

$$= \int_0^t T(s)y\,ds.$$

Because of this and (d) we have

$$\lim_{t\to 0_+} A_t x = \lim_{t\to 0_+} t^{-1}\int_0^t T(s)y\,ds$$

$$= T(0)y = y,$$

which proves that $x \in D(A)$ and $Ax = y$.
   The proof of Theorem 2.2.1 is complete.

   As an application of Theorem 2.2.1, we shall prove that the abstract Cauchy problem

$$dx/dt = Ax, \quad t \geqslant 0 \qquad \text{and} \qquad x(0) = x_0, \quad x_0 \in D(A) \quad (2.2.1)$$

has a unique solution.

THEOREM 2.2.2.   Let A on $D(A)$ be the infinitesimal generator of a strongly continuous semigroup $\{T(t)\}$, $t \geqslant 0$. Then the Cauchy problem (2.2.1) has the unique solution

$$x(t) = T(t)x_0, \qquad t \geqslant 0.$$

Proof: The existence is a consequence of Theorem 2.2.1(b). In fact

$$(d/dt)x(t) = (d/dt)T(t)x_0$$
$$= AT(t)x_0$$
$$= Ax(t)$$

and

$$x(0) = T(0)x_0 = x_0.$$

To prove uniqueness let $y(t)$ be any solution of (2.2.1). Set $F(s) = T(t-s)y(s)$. Since $y(s) \in D(A)$, it follows by Theorem 2.2.1 that the function $F(s)$ is (strongly) differentiable in s and

$$(d/ds)F(s) = -AT(t-s)y(s) + T(t-s)y'(s)$$
$$= -AT(t-s)y(s) + T(t-s)Ay(s) \equiv 0, \qquad 0 \le s \le t.$$

Hence $F(s) = \text{const}$ for $0 \le s \le t$. In particular $F(0) = F(t)$. Since $F(0) = T(t)y(0) = T(t)x_0 = x(t)$ and $F(t) = T(0)y(t) = y(t)$, the proof is complete.

PROBLEM 2.2.2.   If in addition to the hypotheses of Theorem 2.2.2 A generates a contraction semigroup, prove that the norm $\|x(t)\|$ of the solution of (2.2.1) is nonincreasing in t for $t \ge 0$.

As a further application of Theorems 2.2.1 we may consider the non-homogeneous Cauchy problem

$$dx/dt - Ax = f(t), \quad t \ge 0 \qquad \text{and} \qquad x(0) = x_0, \quad x_0 \in D(A). \quad (2.2.2)$$

THEOREM 2.2.3.   Let A on $D(A)$ be the infinitesimal generator of a strongly continuous semigroup $\{T(t)\}$, $t \ge 0$. Let $f: [0, \infty) \to X$ be a (strongly) continuously differentiable function. Then the Cauchy problem (2.2.2) has the unique solution

$$x(t) = T(t)x_0 + \int_0^t T(t-s)f(s)\,ds, \qquad t \ge 0. \qquad (2.2.3)$$

Proof: The uniqueness part of the proof is a consequence of Theorem 2.2.2. For the existence part it suffices to show that the function $x(t)$ in (2.2.3) has a (strong) derivative and satisfies (2.2.2). Obviously $x(0) = x_0$.

Define the function

$$g(t) = \int_0^t T(t-s)f(s)\, ds$$

$$= \int_0^t T(s)f(t-s)\, ds.$$

Since $T(t)$ is bounded for each $t \geqslant 0$ and $f(s)$ is continuous, it follows from Theorem 1.3.1 that the Riemann integral $\int_0^t T(s)f(t-s)\, ds$ exists. We shall first prove that $g(t)$ is (strongly) differentiable. In fact

$$[g(t+h) - g(t)]/h = h^{-1}\int_0^{t+h} T(s)f(t+h-s)\, ds - h^{-1}\int_0^t T(s)f(t-s)\, ds$$

$$= \int_0^t T(s)[f(t+h-s) - f(t-h)]/h\, ds$$

$$+ h^{-1}\int_t^{t+h} T(s)f(t+h-s)\, ds.$$

Hence $g'(t)$ exists and

$$g'(t) = \int_0^t T(s)f'(t-s)\, ds + T(t)f(0).$$

(Although this formula for $g'(t)$ is not needed and only its existence is required, it is of interest to compare it with the classical formula for $g'(t)$.)

On the other hand, for $h > 0$ we have

$$[g(t+h) - g(t)]/h = h^{-1}\int_0^{t+h} T(t+h-s)f(s)\, ds - h^{-1}\int_0^t T(t-s)f(s)\, ds$$

$$= [T(h) - I]/h \int_0^t T(t-s)f(s)\, ds$$

$$+ h^{-1}\int_t^{t+h} T(t+h-s)f(s)\, ds.$$

Since the limit on the left-hand side exists and also $\lim_{h\to 0}\int_t^{t+h} T(t+h-s)$ $\times f(s)\, ds = f(t)$, it follows that $\lim_{h\to 0_+} Ah \int_0^t T(t-s)f(s)\, ds$ exists and is equal to $A \int_0^t T(t-s)f(s)\, ds$. Hence

$$g'(t) = A \int_0^t T(t-s)f(s)\, ds + f(t).$$

As a result of (2.2.3) we get

$$dx(t)/dt = AT(t)x_0 + A \int_0^t T(t-s)f(s)\,ds + f(t)$$
$$= Ax(t) + f(t),$$

and the proof is complete.

REMARK 2.2.1. The conclusion of Theorem 2.2.3 remains valid if, instead of assuming that $f(t)$ is continuously differentiable, we assume that $f(t) \in D(A)$ for all $t \geqslant 0$, and $f(t)$ and $Af(t)$ are strongly continuous in t.

Theorem 2.2.2 can be used to prove the following:

THEOREM 2.2.4. An operator A with domain $D(A)$ dense in the Banach space X can be the infinitesimal generator of at most one strongly continuous semigroup $\{T(t)\}$, $t \geqslant 0$.

Proof: Let $\{T(t)\}$, $t \geqslant 0$, and $\{S(t)\}$, $t \geqslant 0$, be two strongly continuous semigroups of operators in X having A as infinitesimal generator. Consider the Cauchy problem (2.2.1). In view of Theorem 2.2.2 $x(t) = T(t)x_0$ and $y(t) = S(t)x_0$ are solutions of (2.2.1). By uniqueness, it follows that

$$T(t)x_0 = S(t)x_0, \qquad t \geqslant 0, \quad x_0 \in D(A).$$

Since $D(A)$ is dense in X and the operators $T(t)$ and $T(s)$ are bounded, we conclude that

$$T(t)x = S(t)x, \qquad t \geqslant 0, \quad x \in X,$$

which is the desired result.

PROBLEM 2.2.3. Solve the initial value problems

(i) $[\partial u(\xi, t)/\partial t] + \xi^2\,\partial u(\xi, t)/\partial \xi = 0, \qquad t \geqslant 0, \quad 0 \leqslant \xi \leqslant 1,$

where $\lim_{t \to 0} u(\xi, t) = x_0(\xi)$, uniformly in ξ for sufficiently smooth $x_0(\xi)$;

(ii) $\partial u(\xi, t)/\partial t = \partial u(\xi, t)/\partial \xi, \qquad t \geqslant 0, \quad \xi \geqslant 0,$

where $\lim_{t \to 0} u(\xi, t) = x_0(\xi)$, uniformly in ξ for sufficiently smooth $x_0(\xi)$.

[Hint: Use Problem 2.2.1 and Theorem 2.2.2.]

2.3. The Hille–Yosida–Phillips Theorem

Here we shall give a necessary and sufficient condition that an operator A with domain $D(A)$ in the Banach space X is the generator of a strongly continuous semigroup $\{T(t)\}$, $t \geqslant 0$. Recall that the *resolvent set* $\rho(A)$ of A consists of all complex numbers λ for which $(\lambda I - A)^{-1}$ exists as a bounded operator with domain X. The set $\rho(A)$ is an open set in the complex plane C. The function $R(\lambda; A) = (\lambda I - A)^{-1}$, defined on $\rho(A)$ is called the *resolvent* of A and is an analytic function of λ for $\lambda \in \rho(A)$.

THEOREM 2.3.1.   A necessary and sufficient condition that a closed operator A with dense domain $D(A)$ in the Banach space X be the infinitesimal generator of a strongly continuous semigroup $\{T(t)\}$, $t \geqslant 0$, is that there exist real numbers M and ω such that for every real number $\lambda > \omega$

$$\lambda \in \rho(A) \qquad \text{and} \qquad \|R(\lambda; A)^n\| \leqslant M/(\lambda - \omega)^n, \qquad n = 1, 2, \ldots.$$

Proof: We first prove the *sufficiency*. Define the family of bounded operators

$$B_\lambda = \lambda[\lambda R(\lambda; A) - I], \qquad \lambda > \omega.$$

We shall construct the operator $T(t)$ as the strong limit as $\lambda \to \infty$ of the operator $S_\lambda(t)$ where $S_\lambda(t) = \exp(tB_\lambda)$. For convenience we break up the proof into a series of interesting claims.

Claim 1:   The operators $S_\lambda(t)$ are uniformly bounded for λ sufficiently large. In fact, using the Cauchy product of series, we have

$$S_\lambda(t) = \sum_{n=0}^{\infty} \frac{(tB_\lambda)^n}{n!}$$

$$= \sum_{n=0}^{\infty} \frac{(t\lambda^2)^n}{n!} \left[R(\lambda, A) - \frac{I}{\lambda} \right]^n$$

$$= \sum_{n=0}^{\infty} \left\{ 1 \cdot \frac{[t\lambda^2 R(\lambda; A)]^n}{n!} - \frac{\lambda t}{1!} \cdot \frac{[t\lambda^2 R(\lambda; A)]^{n-1}}{(n-1)!} + \cdots \right.$$

$$\left. + (-1)^n \frac{(t\lambda)^n}{n!} \cdot 1 \right\}$$

$$= \sum_{n=0}^{\infty} \frac{(-\lambda t)^n}{n!} \sum_{n=0}^{\infty} \frac{(\lambda^2 t)^n}{n!} R^n(\lambda; A)$$

$$= \exp(-\lambda t) \sum_{n=0}^{\infty} \frac{(\lambda^2 t)^n}{n!} R^n(\lambda; A).$$

Consequently, we obtain

$$\|S_\lambda(t)\| \leqslant \exp(-\lambda t) \sum_{n=0}^\infty [(\lambda^2 t)^n/n!] \|R^n(\lambda;A)\|$$

$$\leqslant \exp(-\lambda t) \sum_{n=0}^\infty M(\lambda^2 t)^n(\lambda-\omega)^{-n}/n!$$

$$= M\exp(-\lambda t)\exp[\lambda^2 t/(\lambda-\omega)]$$

$$= M\exp[\lambda t\omega/(\lambda-\omega)].$$

Since $\lambda t\omega/(\lambda-\omega)\to t\omega$ as $\lambda\to\infty$, it follows that for a fixed $\omega_1 > \omega$ $\|S_\lambda(t)\| \leqslant M\exp(t\omega_1)$ for λ sufficiently large.

Claim 2:

$$\lim_{\lambda\to\infty} \lambda R(\lambda;A)x = x, \qquad x \in X.$$

Notice that

$$\|\lambda R(\lambda;A)\| \leqslant M|\lambda|/(\lambda-\omega) \to M \qquad \text{as} \quad \lambda \to \infty$$

and

$$\|\lambda R(\lambda;A)\| \leqslant 2M \qquad \text{for} \quad \lambda \text{ sufficiently large.}$$

Also for $x \in D(A)$

$$\|\lambda R(\lambda;A)x - x\| = \|R(\lambda;A)Ax\| \leqslant M/(\lambda-\omega)\|Ax\| \to 0 \qquad \text{as} \quad \lambda \to \infty$$

and Claim 2 follows from Appendix VI.

Claim 3:

$$\lim_{\lambda\to\infty} B_\lambda x = Ax, \qquad x \in D(A).$$

This is evident from Claim 2, because

$$\lim_{\lambda\to\infty} B_\lambda x = \lim_{\lambda\to\infty} \lambda R(\lambda;A)Ax = Ax.$$

Claim 4: For every $t \geqslant 0$ and $x \in X$ we have that the $\lim_{\lambda\to\infty} S_\lambda(t)x$ exists and defines a bounded operator $T(t)$. Set $R_\lambda = (\lambda I - A)^{-1}$ for $\lambda \in \rho(A)$. From Appendix VIII $R_\lambda R_\mu = R_\mu R_\lambda$. Therefore $B_\lambda B_\mu = B_\mu B_\lambda$. Since $S_\lambda(t) = \sum_{n=0}^\infty (t^n/n!)B_\lambda^n$, we then conclude that $B_\mu S_\lambda(t) = S_\lambda(t)B_\mu$ for $\lambda, \mu \in \rho(A)$.

By Theorem 1.3.4, we have for every $x \in X$

$$S_\lambda(t)x - S_\mu(t)x = \int_0^t (d/ds)[S_\mu(t-s)S_\lambda(s)x]\,ds$$

$$= \int_0^t [-S_\mu(t-s)B_\mu S_\lambda(s) + S_\mu(t-s)S_\lambda(s)B_\lambda]x\,ds$$

$$= \int_0^t S_\mu(t-s)S_\lambda(s)(B_\lambda - B_\mu)x\,ds.$$

Using Claims 1 and 3, we obtain for $x \in D(A)$

$$\|S_\lambda(t)x - S_\mu(t)x\| \leqslant \int_0^t M\exp[(t-s)\omega_1]\,M\exp(s\omega_1)\,\|B_\lambda x - B_\mu x\|\,ds$$

$$= M^2 \exp(t\omega_1)\,\|B_\lambda x - B_\mu x\| \to 0 \qquad \text{as} \quad \lambda, \mu \to \infty.$$

Moreover, the convergence is uniform in every finite interval of t. This together with Appendix VIII shows that there exists a bounded operator $T(t)$ such that

$$\lim_{\lambda \to \infty} S_\lambda(t)x = T(t)x, \qquad x \in X. \tag{2.3.1}$$

Clearly $\|T(t)\| \leqslant M\exp(t\omega_1)$ and Claim 4 is established.

Claim 5: The family $\{T(t)\}$, $t \geqslant 0$, is a strongly continuous semigroup in X.

Indeed, the properties (i) and (ii) of Definition 2.1.1 follow directly from (2.3.1) and the fact that the operators $S_\lambda(t)$ themselves satisfy these properties. As the convergence (2.3.1) is uniform in every finite interval of t, the limit $T(t)x$ is strongly continuous in t for each $x \in X$, and property (iii) of semigroups is established.

To complete the proof of the sufficiency we only have to show that A is the infinitesimal generator of $\{T(t)\}$. Let B be the infinitesimal generator of the semigroup $\{T(t)\}$, $t \geqslant 0$. We must prove that $A = B$. We first derive a formula for the resolvent $R(\lambda; B)$ of the infinitesimal generator B of $\{T(t)\}$ where $\|T(t)\| \leqslant M\exp(t\omega_1)$ for $t \geqslant 0$.

Claim 6: Every $\lambda > \omega_1$ is in $\rho(B)$ and

$$R(\lambda; B)x = \int_0^\infty \exp(-\lambda t)T(t)x\,dt, \qquad x \in X, \quad \lambda > \omega_1. \tag{2.3.2}$$

Consider the operator $R(\lambda)$ defined by

$$R(\lambda)x \equiv \int_0^\infty \exp(-\lambda t)\,T(t)\,x\,dt, \qquad x \in X, \quad \lambda > \omega_1. \qquad (2.3.3)$$

Since

$$\|\exp(-\lambda t)\,T(t)\| \leqslant \exp(-\lambda t)\,M\exp(t\omega_1)$$
$$= M\exp[-(\lambda - \omega_1)t],$$

the integral in (2.3.3) is absolutely convergent for $\lambda > \omega_1$ (in fact for $\mathrm{Re}(\lambda) > \omega_1$) and defines a bounded operator $R(\lambda)$ in X.

Observe that, for $h > 0$ and $x \in X$

$$B_h R(\lambda)x = [T(h)R(\lambda)x - R(\lambda)x]/h$$

$$= h^{-1}\int_0^\infty \exp(-\lambda t)\,T(t+h)\,x\,dt - h^{-1}\int_0^\infty \exp(-\lambda t)\,T(t)\,x\,dt$$

$$= h^{-1}\int_h^\infty \exp[-\lambda(s-h)]\,T(s)\,x\,ds - h^{-1}\int_0^\infty \exp(-\lambda t)\,T(t)\,x\,dt$$

$$= [\exp(\lambda h)-1]/h\int_0^\infty \exp(-\lambda t)\,T(t)\,x\,dt$$

$$- \exp(\lambda h)\,h^{-1}\int_0^h \exp(-\lambda t)\,T(t)\,x\,dt.$$

Hence

$$\lim_{h\to 0_+} B_h R(\lambda)x = R(\lambda)x - x.$$

This proves that $R(\lambda)x \in D(B)$ and $BR(\lambda)x = \lambda R(\lambda)x - x$, that is,

$$(\lambda I - B)\,R(\lambda)x = x, \qquad x \in X, \quad \lambda > \omega_1. \qquad (2.3.4)$$

On the other hand, since B is a closed operator we have from Theorem 1.3.5 for any $x \in D(B)$

$$BR(\lambda)x = B\int_0^\infty \exp(-\lambda t)\,T(t)\,x\,dt$$

$$= \int_0^\infty \exp(-\lambda t)\,T(t)\,Bx\,dt$$

$$= R(\lambda)\,Bx.$$

In view of (2.3.4) we then have

$$R(\lambda)(\lambda I - B)x = x, \qquad x \in D(A), \quad \lambda > \omega_1. \qquad (2.3.5)$$

The relations (2.3.4) and (2.3.5) prove Claim 6.

Claim 7: $D(A) \subset D(B)$ and $Ax = Bx$ for $x \in D(A)$. For $x \in X$ we have the identity

$$S_\lambda(t)x - x = \int_0^t (d/ds)[S_\lambda(s)x]\,ds$$

$$= \int_0^t S_\lambda(s)B_\lambda x\,ds. \qquad (2.3.6)$$

We would like to take limits as $\lambda \to \infty$ on both sides of (2.3.6). To this end observe that for $x \in D(A)$

$$\|S_\lambda(s)B_\lambda x - T(s)Ax\| \leqslant \|S_\lambda(s)\|\,\|B_\lambda x - Ax\| + \|[S_\lambda(s) - T(s)]Ax\|$$

$$\leqslant M\exp(\omega_1 s)\|B_\lambda x - Ax\|$$

$$+ 2M\exp(\omega_1 s)\|Ax\| \to 0 \qquad \text{as} \quad \lambda \to \infty,$$

uniformly in S on every closed interval $[0, t]$. We conclude that

$$T(t)x - x = \int_0^t T(s)Ax\,ds, \qquad x \in D(A).$$

Hence

$$Bx = \lim_{t \to 0_+}[T(t)x - x]/t$$

$$= \lim_{t \to 0_+} t^{-1}\int_0^t T(s)Ax\,ds$$

$$= Ax,$$

and Claim 7 is proved.

Claim 8: $D(B) \subset D(A)$. By Claim 6, we have $\rho(A) \cap \rho(B) \neq \varnothing$. Let $\lambda_0 \in \rho(A) \cap \rho(B)$. Then by Claim 7

$$(\lambda_0 I - B)D(A) = (\lambda_0 I - A)D(A)$$

$$= X$$

$$= (\lambda_0 I - B)D(B).$$

So if $x_\beta \in D(B)$, there exists an $x_\alpha \in D(A)$ such that

$$(\lambda_0 I - B) x_\alpha = (\lambda_0 I - B) x_\beta.$$

Since $\lambda_0 I - B$ is one to one, we obtain $x_\beta = x_\alpha \in D(A)$ and the claim is proved.

The proof of the sufficiency of Theorem 2.3.1 is therefore complete.

Next we prove there exist real numbers M and ω such that

$$\|T(t)\| \leq M \exp(\omega t), \qquad t \geq 0.$$

As proved in Claim 6, every $\lambda > \omega$ is in $\rho(A)$ and for $\lambda > \omega$

$$R(\lambda; A) x = \int_0^\infty \exp(-\lambda t) T(t) x \, dt, \qquad x \in X. \tag{2.3.7}$$

From the resolvent formula (see Appendix VIII) one obtains

$$R(\lambda; A) - R(\mu; A) = (\mu - \lambda) R(\lambda; A) R(\mu, A), \qquad \lambda, \mu > \omega.$$

Consequently the analyticity of $R(\lambda; A)$ for $\lambda \in \rho(A)$ yields

$$(d/d\lambda) R(\lambda; A) = \lim_{\mu \to \lambda} [R(\lambda; A) - R(\mu; A)]/(\lambda - \mu)$$
$$= -R(\lambda; A)^2.$$

It therefore follows by induction

$$(d^n/d\lambda^n) R(\lambda; A) = (-1)^n n! R(\lambda; A)^{n+1}, \qquad \lambda > \omega. \tag{2.3.8}$$

On the other hand, differentiation with respect to λ under the integral sign of (2.3.7) is justifiable. In fact

$$([R(\lambda + h; A) - R(\lambda; A)]/h) x$$
$$= h^{-1} \int_0^\infty ([\exp(-ht) - 1]/h) \exp(-\lambda t) T(t) x \, dt. \tag{2.3.9}$$

Since

$$([\exp(-ht) - 1]/h) \exp(-\lambda t) T(t) x \to -t \exp(-\lambda t) T(t) x \qquad \text{as} \quad h \to 0$$

and

$$\|([\exp(-ht) - 1]/h) \exp(-\lambda t) T(t) x\| \leq Mt \exp[-(\lambda - \omega - |h|) t],$$

we can take limits as $h \to 0$ is (2.3.9) getting

$$(d/d\lambda) R(\lambda; A) x = \int_0^\infty -t \exp(-\lambda t) T(t) x \, dt.$$

Thus one easily gets after n steps

$$(d^n/d\lambda^n)[R(\lambda;A)x] = (-1)^n \int_0^\infty t^n \exp(-\lambda t) T(t) x \, dt, \qquad x \in X, \quad \lambda > \omega.$$
(2.3.10)

Comparing (2.3.8) and (2.3.10), there results

$$R(\lambda;A)^n = [(n-1)!]^{-1} \int_0^\infty t^{n-1} \exp(-\lambda t) T(t) x \, dt. \qquad (2.3.11)$$

Hence

$$\|R(\lambda;A)^n\| \leqslant M/(n-1)! \int_0^\infty t^{n-1} \exp[-(\lambda-\omega)t] \, dt$$

$$= M/(\lambda-\omega)^n, \qquad \lambda > \omega$$

and the proof is complete.

PROBLEM 2.3.1. Let A be a linear operator in X with domain $D(A)$ dense in X. The following statements are equivalent:

(i) A generates a strongly continuous semigroup $\{T(t)\}$, $t \geqslant 0$ such that $\|T(t)\| \leqslant \exp(\omega t)$ for some real number ω;
(ii) there is a real number ω such that for $\lambda > \omega$

$$\lambda \in \rho(A) \qquad \text{and} \qquad \|R(\lambda;A)\| \leqslant (\lambda-\omega)^{-1}.$$

[Hint: Use Theorem 2.3.1 and the estimate $\|R(\lambda;A)^n\| \leqslant \|R(\lambda;A)\|^n$.]

PROBLEM 2.3.2. Prove Theorem 2.2.4 by utilizing formula (2.3.7).

PROBLEM 2.3.3. For the semigroup of Example 2.1.3 show that

$$\rho(A) = \{\lambda \in C: R(\lambda) > 0\}$$

and

$$R(\lambda;A)x(t) = \int_0^\infty \exp(-\lambda s) x(t+s) \, ds, \qquad R(\lambda) > 0.$$

[Hint: Consider the general solution of the equation $\lambda y - y' = x$.]

As an application of Theorem 2.3.1 we prove a uniqueness result for the solutions of the abstract Cauchy problem

$$dx/dt = A(t)x, \qquad a \leqslant t \leqslant b; \qquad (2.3.12)$$

$$x(a) = x_a, \qquad x_a \in D[A(a)]. \qquad (2.3.13)$$

We need the following:

LEMMA 2.3.1.   Let A on $D(A)$ be the infinitesimal generator of a contraction semigroup. Then for each $x \in D(A)$

$$\|(I+\varepsilon A)x\| \leqslant \|x\| + 0(\varepsilon) \qquad \text{as} \quad \varepsilon \to 0_+ . \qquad (2.3.14)$$

Proof: By Theorem 2.3.1 and the hypothesis that A generates a contraction semigroup, it follows that $(0, \infty) \subset \rho(A)$ and $\|(\lambda I - A)^{-1}\| \leqslant \lambda^{-1}$ for $\lambda > 0$. Hence, for $\varepsilon = \lambda^{-1} > 0$ we have the estimate

$$\|(I-\varepsilon A)^{-1}\| \leqslant 1, \qquad \varepsilon > 0. \qquad (2.3.15)$$

Observe that for $x \in D(A)$ and $\varepsilon > 0$

$$
\begin{aligned}
(I+\varepsilon A)x &= (I+\varepsilon A)(I-\varepsilon A)(I-\varepsilon A)^{-1}x \\
&= (I-\varepsilon^2 A^2)(I-\varepsilon A)^{-1}x \\
&= (I-\varepsilon A)^{-1}x - \varepsilon^2 A^2 (I-\varepsilon A)^{-1}x \\
&= (I-\varepsilon A)^{-1}x - \varepsilon B_\varepsilon Ax \qquad (2.3.16)
\end{aligned}
$$

where $B_\varepsilon = A(I-\varepsilon A)^{-1} = (I-\varepsilon A)^{-1} - I$. So $\|B_\varepsilon\| \leqslant 2$. For any $y \in D(A)$ we have

$$
\begin{aligned}
\|B_\varepsilon y\| &= \|\varepsilon A (I-\varepsilon A)^{-1}y\| \\
&\leqslant \varepsilon \|Ay\| \to 0.
\end{aligned}
$$

From Appendix VI it follows that for each $x \in X$ we have $\lim_{\varepsilon \to 0_+} B_\varepsilon x = 0$. In view of (2.3.16) and (2.3.15) we have

$$
\begin{aligned}
\|(I+\varepsilon A)x\| &\leqslant \|x\| + \varepsilon \|B_\varepsilon Ax\| \\
&= \|x\| + O(\varepsilon)
\end{aligned}
$$

and the proof is complete.

THEOREM 2.3.2.   Assume that

(i)   for each $t \in [a, b]$, the operator $A(t)$ with domain $D[A(t)]$ generates a contraction semigroup in the Banach space X;

(ii) $x \in C[[a, b], X]$, $x(t) \in D[A(t)]$ and has a strongly continuous, right derivative $x_+'(t)$ such that

$$x_+'(t) = A(t)x(t), \qquad a \leqslant t \leqslant b.$$

Then $\|x(t)\|$ is nonincreasing in t for $t \in [a, b]$.

Proof: From the definition of $x_+'(t)$ we have

$$\|x(t+\varepsilon) - [I+\varepsilon A(t)]x(t)\| = O(\varepsilon) \quad \text{as} \quad \varepsilon \to 0_+.$$

This and Lemma 2.3.1 yield the inequality

$$\|x(t+\varepsilon)\| \leqslant \|[I+\varepsilon A(t)]x(t)\| + O(\varepsilon)$$

$$\leqslant \|x(t)\| + O(\varepsilon).$$

Hence $\|x(t)\|_+' \leqslant 0$ for $a \leqslant t \leqslant b$. Since by hypothesis $x(t)$ is continuous, the result follows.

COROLLARY 2.3.1. Under the hypothesis (i) of Theorem 2.3.2 the abstract Cauchy problem (2.3.12) and (2.3.13) has at most one solution on $[a, b]$.

Proof: Let $x_1(t)$ and $x_2(t)$ be two solutions of the system (2.3.12) and (2.3.13). Set $x(t) = x_1(t) - x_2(t)$. Then $x'(t) = A(t)x(t)$ and $x(a) = 0$. From Theorem 2.3.2 $\|x(t)\|$ is nonincreasing in t. Since $x(a) = 0$, it follows that $x(t) \equiv 0$, and the proof is complete.

We shall present another application of Theorem 2.3.1. Let $X = C_0(-\infty, \infty)$ be the space of continuous complex-valued functions which tend to zero at infinity. Consider the heat equation

$$\partial u/\partial t = \partial^2 u/\partial x^2 \quad \text{and} \quad u(0, x) = u_0(x), \quad -\infty < x < \infty$$

Let $D(A)$ be the class of functions $y(x)$ with y and dy/dx continuously differentiable and y, d^2y/dx^2 in X. Define the operator A with domain $D(A)$ by

$$Ay = d^2y/dx^2.$$

Clearly, $D(A)$ is dense in X and A is a closed operator on $D(A)$. A solution of

$$(\lambda I - A)y = \lambda y - d^2y/dx^2$$

$$= u_0(x)$$

in $D(A)$ can be obtained, by the method of variation of parameters, in the form

$$y(x) = (2\sqrt{\lambda})^{-1} \int_{-\infty}^{\infty} \exp[-\sqrt{\lambda}(x-s)] u_0(s) \, ds.$$

Using the integral representation of $R(\lambda; A)$ [18], it follows that $\|R(\lambda; A)\| \leqslant \lambda^{-1}$ for $\lambda > 0$. Therefore A on $D(A)$ generates a contraction semigroup in X.

2.4. Linear Autonomous Functional Differential Equations

The theory of strongly continuous semigroups finds an interesting application in linear autonomous functional differential equations. Let $\tau > 0$ be given. Let $X = C[[-\tau, 0], R^n]$ denote the Banach space of continuous functions with domain $[-\tau, 0]$ and range in R^n with the norm of $\phi \in X$ defined by

$$\|\phi\| = \sup_{-\tau \leqslant s \leqslant 0} |\phi(s)| \qquad (|\cdot| \text{ any norm in } R^n).$$

Suppose that $x \in C[[-\tau, \infty), R^n]$. For any $t \geqslant 0$, we shall let x_t denote the element of X defined by $x_t(s) = x(t+s)$ for $-\tau \leqslant s \leqslant 0$. x_t is called the *past history* of x at t. Let $L: X \to R^n$ be a continuous (and so bounded) linear operator mapping X into R^n. By Riesz representation theorem (see Appendix IV), there exists an $n \times n$ matrix $n(\theta)$, the elements of which are of bounded variation such that

$$L(\phi) = \int_{-\tau}^{0} [dn(\theta)]\,\phi(\theta), \qquad \phi \in X \qquad \text{(Stieltjes integral)}.$$

With this notation consider the linear autonomous functional differential equation

$$\dot{x}(t) = L(x_t). \tag{2.4.1}$$

It is known (see [42]) that for any $\phi_0 \in X$, (2.4.1) has a unique solution with initial function ϕ_0 at $t = t_0$, that is, the past history of the solution at t_0 is ϕ_0. The solution is denoted by $x(t_0, \phi_0)$ and satisfies (2.4.1) for all $t \geqslant t_0$. If $t_0 = 0$, we usually set $x(0, \phi_0) = x(\phi_0)$. For $t \geqslant 0$ we define the operator $T(t): X \to X$ by the relation $T(t)\phi = x_t(\phi)$ for $\phi \in X$.

THEOREM 2.4.1.   (a)   For each fixed $t \geqslant 0$ the operator $T(t)$ is linear and bounded and for $t \geqslant \tau$, $\{T(t)\}$ is completely continuous (compact);

(b)   the family $\{T(t)\}$, $t \geqslant 0$ is a strongly continuous semigroup on X;

(c)   the infinitesimal generator A of this semigroup has for domain $D(A)$ the space $C_1[-\tau, 0]$ of continuously differentiable functions on $[-\tau, 0]$ with $\dot{\phi}(0-) = L(\phi)$ and $A = d/d\theta$ on $D(A)$, that is,

$$A\phi(\theta) = d\phi(\theta)/d\theta, \qquad -\tau \leqslant \theta \leqslant 0 \quad \text{and} \quad A\phi(0-) = L(\phi).$$

Proof: (a)   (i) $T(t): X \to X$ is linear. Let $\phi, \psi \in X$ and λ, μ be any two scalars. Since L is linear, $\lambda x(\phi) + \mu x(\psi)$ is a solution of (2.4.1) with initial function $\lambda\phi + \mu\psi$ at $t = 0$. By the uniqueness of solutions of (2.4.1) we have the equality

$$x(\lambda\phi + \mu\psi) = \lambda x(\phi) + \mu x(\psi), \qquad t \geqslant 0.$$

Hence

$$x_t(\lambda\phi + \mu\psi) = \lambda x_t(\phi) + \mu x_t(\psi),$$

that is,

$$T(t)(\lambda\phi + \mu\psi) = \lambda T(t)\phi + \mu T(t)\psi.$$

(ii) $T(t): X \to X$ is bounded. Since $L: X \to R^n$ is a continuous linear operator it is also bounded, that is, there exists an $l > 0$ such that $|L(\phi)| < l\|\phi\|$ for each $\phi \in X$. From the definition of $T(t)$ and $x_t(\phi)$ it follows that

$$T(t)\phi(0) = x(\phi)(t+\theta),$$

and so integrating (2.4.1) we get for $-\tau \leqslant \theta \leqslant 0$

$$T(t)\phi(0) = \phi(t+\theta), \qquad\qquad t + \theta \leqslant 0$$

$$= \phi(0) + \int_0^{t+\theta} L[T(s)\phi]\, ds, \qquad t + \theta \geqslant 0.$$

It follows that

$$|T(t)\phi(0)| \leqslant \|\phi\| + \int_0^t l\|T(s)\phi\|\, ds;$$

so

$$\|T(t)\phi\| \leqslant \|\phi\| + \int_0^t l\|T(s)\phi\|\, ds.$$

From Gronwall's inequality we conclude that

$$\|T(t)\phi\| \leqslant \exp(lt)\|\phi\|, \qquad t \geqslant 0, \quad \phi \in X.$$

Hence $T(t)$ is bounded by $\exp(lt)$.

(iii) $T(t): X \to X$ for $t \geqslant \tau$ is completely continuous. If S is the closed unit sphere $S = \{\phi \in X: \|\phi\| \leqslant 1\}$, we must show that the strong closure of $\{T(t)S\}$ is compact in the strong topology of X. It suffices to show that $\{T(t)S\}$ is a family of uniformly bounded and equicontinuous functions on $[-\tau, 0]$. Indeed, let $\psi = T(t)\phi$ for $\phi \in S$. Then $\|\psi\| = \|T(t)\phi\| \leqslant \exp(lt)\|\phi\| \leqslant \exp(lt)$ which proves that the family $\{T(t)S\}$ is uniformly bounded. Next we show that the family of functions $\{T(t)S\}$ is equicontinuous by showing that their derivatives are uniformly bounded on $[-\tau, 0]$. In fact, let $\psi = T(t)\phi$ for $\phi \in S$. Then

$$\psi(\theta) = T(t)\phi(0) = x(\phi)(t+\theta), \qquad -\tau \leqslant \theta \leqslant 0.$$

For $t \geqslant \tau$ we have $t + \theta \geqslant 0$; therefore

$$\dot{\psi}(\theta) = \dot{x}(\phi)(t+\theta) = L[x_{t+\theta}(\phi)].$$

It follows that

$$|\dot{\psi}(\theta)| = |L(T(t+\theta)\phi)|$$
$$\leqslant l\|T(t+\theta)\phi\|$$
$$\leqslant l\exp[l(t+\theta)]\|\phi\|$$
$$\leqslant l\exp(lt),$$

which proves our assertion.

(b) (i) $\qquad\qquad T(t)\,T(s) = T(t+s), \qquad t,s \geqslant 0.$

We must show that for any $\phi \in X$,

$$T(t)\,T(s)\,\phi = T(t+s)\,\phi, \qquad \text{that is} \qquad x_t[x_s(\phi)] = x_{t+s}(\phi),$$

which is true on the strength of uniqueness of (2.4.1).

(ii) $\qquad T(0)\,\phi = x_0(\phi) = \phi, \qquad \phi \in X, \qquad \text{that is} \qquad T(0) = I.$

(iii)   For each $\phi \in X$ we have that $T(t)\,\phi$ is strongly continuous in t for $t \geqslant 0$. In fact

$$[T(t+\Delta t)\,\phi - T(t)\,\phi]\,\theta = [x_{t+\Delta t}(\phi) - x_t(\phi)]\,\theta$$
$$= x(\phi)(t+\Delta t+\theta) - x(\phi)(t+\theta).$$

Therefore

$$[T(t+\Delta t)\,\phi - T(t)\,\phi]\,\theta = \int_{t+\theta}^{t+\Delta t+\theta} L[T(s)\,\phi]\,ds,$$

$$t + \theta > 0, \quad \Delta t \text{ small}$$

$$= \phi(t+\Delta t+\theta) - \phi(t+\theta),$$

$$t + \theta < 0, \quad \Delta t \text{ small}$$

$$= \int_0^{\Delta t} L(T(s)\,\phi)\,ds,$$

$$t + \theta = 0, \quad \Delta t > 0 \text{ and small}$$

$$= \phi(\Delta t) - \phi(0),$$

$$t + \theta = 0, \quad \Delta t < 0 \text{ and small}.$$

In any case, we conclude that $\| T(t+\Delta t)\phi - T(t)\phi \| \to 0$ as $\Delta t \to 0$.

(c) Let A be the infinitesimal generator of the semigroup $\{T(t)\}$ with domain $D(A)$. Then

$$D(A) = \{\phi \in X : \lim_{t \to 0_+} [T(t)\phi - \phi]/t \quad \text{exists}\}.$$

Define also the set

$$\tilde{D} = \{\phi \in X : \phi \in C_1[-\tau, 0] \quad \text{with} \quad \dot\phi(0-) = L(\phi)\}.$$

We must prove that $D(A) = \tilde{D}$. In fact, let $\phi \in \tilde{D}$. Then

$$|([T(t)\phi - \phi]/t)s - \dot\phi(s)|$$

$$= |([x_t(\phi)(s) - \phi(s)]/t) - \dot\phi(s)|$$

$$= \begin{cases} |([\phi(t+s) - \phi(s)]/t) - \dot\phi(s)|, & t > 0 \text{ and small}, \quad s < 0 \\ ([x(\phi)t - \phi(0)]/t) - \dot\phi(0)|, & s = 0. \end{cases}$$

$$= \begin{cases} \left| t^{-1} \int_s^{t+s} [\dot\phi(\xi) - \dot\phi(s)]\, d\xi \right|, & s < 0 \\ \left| t^{-1} \int_0^t (L[T(\xi)]\phi - L(\phi))\, d\xi \right|, & s = 0. \end{cases}$$

In view of the continuity of $\dot\phi(\xi)$ and Theorem 2.2.1(d) it follows that $\phi \in D(A)$. Conversely, let $\phi \in D(A)$. Then the $\lim_{t \to 0_+} [T(t)\phi - \phi]/t$ exists in the topology of X. Observe that

$$([T(t)\phi - \phi]/t)s = \begin{cases} [x_t(\phi)s - \phi(s)]/t, & t > 0 \text{ and small}, \quad s < 0 \\ [x_t(\phi)0 - \phi(0)]/t, & s = 0 \end{cases}$$

$$= \begin{cases} [\phi(t+s) - \phi(s)]/t, & s < 0 \\ t^{-1} \int_0^t L[T(\xi)\phi]\, d\xi, & s = 0. \end{cases}$$

Therefore

$$\lim_{t \to 0_+} ([T(t)\phi - \phi]/t)s = \begin{cases} \dot\phi(s_+), & s < 0 \\ L(\phi), & s = 0. \end{cases}$$

So $\dot\phi(s_+)$ exists and is continuous. It follows from [6] that $\dot\phi(s)$ exists and is continuous. Also the function $A\phi$ is continuous, and $\dot\phi(0_-)$ exists. Hence, $\phi \in \tilde{D}$.

The proof is complete.

Since $T(t)$ will generally not be known, we hope to discuss most of the properties of $T(t)$ by using only properties of the known operator A. Recall that for any $\phi \in D(A)$

$$(d/dt)\, T(t)\,\phi = AT(t)\,\phi$$

and therefore, the abstract Cauchy problem

$$du/dt = Au, \qquad t \geqslant 0$$

$$u(0) = \phi, \qquad \phi \in D(A)$$

has the unique solution $u(t) = T(t)\,\phi = x_t(\phi)$ for $t \geqslant 0$.

We denote by $\sigma_p(A)$ the point spectrum of A, that is, all those values λ in the spectrum $\sigma(A)$ of A for which $(\lambda I - A)$ is not one to one. The points λ in $\sigma_p(A)$ are called the eigenvalues of A, and any $\phi \in X$ such that $(\lambda I - A)\,\phi = 0$ is called an eigenvector corresponding to the eigenvalue λ. For a given $\lambda \in \sigma_p(A)$ the generalized eigenspace of λ is denoted by $M_\lambda(A)$ and is defined to be the smallest subspace of X containing all the elements of X which belong to the null spaces $N[(\lambda I - A)^k]$ with $k = 1, 2, \ldots$. We have the following:

THEOREM 2.4.2. Let A on $D(A)$ be the infinitesimal generator of the strongly continuous semigroup $\{T(t)\}$, $t \geqslant 0$ associated with the linear autonomous functional differential equation (2.4.1). Then

(a) $\qquad\qquad \sigma(A) = \sigma_p(A) = \{\lambda \in C: \det \Delta(\lambda) = 0\}$ $\qquad$ (2.4.2)

where $\Delta(\lambda) = \lambda I - \int_{-\tau}^{0} [dn(\theta)] \exp(\lambda \theta)$;

(b) the real parts of the eigenvalues of A are bounded above;

(c) $\qquad M_\lambda(A) = N[(A - \lambda I)^k], \qquad k \qquad$ some finite integer;

(d) $\qquad\qquad X = N[(A - \lambda I)^k \oplus R[(A - \lambda I)^k]$

where $R[(A - \lambda I)^k]$ denotes the range of $(A - \lambda I)^k$ and $\oplus$ means the direct sum.

Proof: (a) It suffices to show that the resolvent set $\rho(A)$ consists of all λ except those which satisfy (2.4.2) and then show that any λ satisfying (2.4.2) is in $\sigma_p(A)$. By definition $\lambda \in \rho(A)$ if and only if $(A - \lambda I): D(A) \to X$ is one to one and onto, and $(A - \lambda I)^{-1} \in B(X)$. It follows that $\lambda \in \rho(A)$ if and only if for every $\psi \in X$ the equation

$$(A - \lambda I)\,\phi = \psi$$ $\qquad$ (2.4.3)

has a unique solution $\phi \in D(A)$ that depends continuously upon ψ. Since any $\phi \in D(A)$ must be continuously differentiable and $A\phi = \dot{\phi}$, a solution of (2.4.3) must satisfy the differential equation

$$\dot{\phi}(\theta) - \lambda\phi(\theta) = \psi(\theta), \qquad -\tau \leqslant \theta \leqslant 0.$$

Hence

$$\phi(\theta) = \exp(\lambda\theta)b + \int_0^\theta \exp[\lambda(\theta-\xi)]\psi(\xi)\,d\xi, \qquad -\tau \leqslant \theta \leqslant 0. \quad (2.4.4)$$

But $\phi \in D(A)$ implies

$$\dot{\phi}(0) = L(\phi) = \int_{-\tau}^0 [dn(\theta)]\,\phi(\theta).$$

Therefore

$$\lambda b + \psi(0) = \int_{-\tau}^0 [dn(\theta)]\left[\exp(\lambda\theta)b + \int_0^\theta \exp[\lambda(\theta-\xi)]\psi(\xi)\,d\xi\right].$$

Simplifying this expression, we obtain

$$\Delta(\lambda)b = -\psi(0) + \int_{-\tau}^0 \int_0^\theta \exp[\lambda(\theta-\xi)][dn(\theta)]\psi(\xi)\,d\xi. \quad (2.4.5)$$

In view of (2.4.4) and (2.4.5) we obtain

$$\rho(A) = \{\lambda \in C : \det\Delta(\lambda) \neq 0\},$$

and if $\det\Delta(\lambda) \neq 0$, there exists a nontrivial solution of (2.4.3) for $\psi = 0$, that is, $\lambda \in \sigma_p(A)$.

(b) This can be proved as in Claim 6 of Theorem 2.3.1.

(c) and (d) Since $\det\Delta(\lambda)$ is an entire function of λ, it has zeros of finite order. It follows from (2.4.4) and (2.4.5) that the resolvent function $(\lambda I - A)^{-1}$ has a pole of order k at $\lambda = \lambda_0$, if λ_0 is a zero of $\det\Delta(\lambda)$ of order k. Since A is a closed operator, (c) and (d) follow from [74].

The proof is complete.

From Theorem 2.4.2, we know that if $\lambda \in \sigma(A)$, then $M_\lambda(A)$ is finite dimensional and $M_\lambda(A) = N(A - \lambda I)^k$ for some integer k. Since A commutes with $(A - \lambda I)^k$, it follows that $AM_\lambda(A) \subset M_\lambda(A)$.

Let $\langle \phi_1^\lambda, \phi_2^\lambda, \ldots, \phi_d^\lambda \rangle$ be a basis for $M_\lambda(A)$ and let B_λ be the restriction of A to $M_\lambda(A)$. Then $A\Phi_\lambda = \Phi_\lambda B_\lambda$. The $d \times d$ matrix B_λ has λ as its only eigenvalue. Indeed if $\mu \neq \lambda$ is an eigenvalue of B_λ, there exists a

$w \in N[(A - \lambda I)^k]$ such that $B_\lambda w = \mu w$. So $(B_\lambda - \lambda I) w = (A - \lambda I) w = (\mu - \lambda) w$. This implies that $(A - \lambda I)^{k-1} w = 0$ and inductively $(\mu - \lambda) w = 0$, a contradiction. From the definition of A the relation $A\Phi_\lambda = \Phi_\lambda B_\lambda$ implies that

$$\Phi_\lambda(\theta) = \Phi_\lambda(0) \exp(B_\lambda \theta), \qquad -\tau \leqslant \theta \leqslant 0.$$

Also, from Theorem 2.2.1(b), we obtain

$$T(t) \Phi_\lambda = \Phi_\lambda \exp(B_\lambda t), \qquad t \geqslant 0.$$

Hence

$$[T(t) \Phi_\lambda] \theta = \Phi_\lambda(0) \exp[B_\lambda(t + \theta)], \qquad -\tau \leqslant \theta \leqslant 0.$$

This relation can be used to define $T(t)$ on $M_\lambda(A)$ for all values of t in R. All these observations lead to the interesting conclusion that on generalized eigenspaces (2.4.1) behaves essentially as an ordinary differential equation. The decomposition of X into two subspaces invariant under A and $T(t)$ can be used to introduce a coordinate system in X, which plays the role of the Jordan canonical form in ordinary differential equations.

2.5. Analytic Semigroups

In the sequel we shall denote by $\{\exp(-tA)\}$ the strongly continuous semigroup generated by $-A$. Here we shall introduce an important class of semigroups, namely, strongly continuous semigroups $\{\exp(-tA)\}$ which can, as functions of the parameter t, be continued analytically into a sector of the complex plane C containing the positive t-axis. The symbol S_ω will denote the sector $S_\omega = \{t \in C : |\arg t| < \omega, \ t \neq 0\}$.

DEFINITION 2.5.1.   A strongly continuous semigroup $\{\exp(-tA)\}$, is called an *analytic semigroup* if the following conditions are satisfied:

(i) $\exp(-tA)$ can be continued analytically as a strongly continuous semigroup into a sector S_ω for some $\omega \in (0, \pi/2)$;

(ii)   for each $t \in S_\omega$ the operators $A \exp(-tA)$ and $(d/dt) \exp(-tA)$ are in $B(X)$ and

$$(d/dt) \exp(-tA) x = -A \exp(-tA) x, \qquad x \in X; \qquad (2.5.1)$$

(iii)   for any $0 < \varepsilon < \omega$ the operators $\exp(-tA)$ and $tA \exp(-tA)$ are uniformly bounded in the sector $S_{\omega - \varepsilon}$, that is, there exists a constant

$K = K(\varepsilon)$ such that

$$\|\exp(-tA)\| \leqslant K \quad \text{and} \quad \|A\exp(-tA)\| \leqslant K/|t|, \quad t \in S_{\omega-\varepsilon}.$$

$$(2.5.2)$$

Analytic semigroups arise in parabolic partial differential equations $u_t = -\tau u$ where τ is a formal elliptic partial differential operator satisfying the hypotheses of Garding's inequality [19].

The following theorem gives sufficient conditions for an unbounded operator $-A$ to generate an analytic semigroup.

THEOREM 2.5.1. Assume that

(i) A is a closed operator with domain $D(A)$ dense in X;
(ii) the resolvent set $\rho(-A)$ of $-A$ contains the sector $S_{(\pi/2)+\omega}$ for some $\omega \in (0, \pi/2)$;
(iii) the resolvent functions of $-A$ satisfies

$$\|(\lambda I + A)^{-1}\| \leqslant M/|\lambda|, \quad \lambda \in S_{(\pi/2)+\omega}, \quad (2.5.3)$$

where M is a constant independent of λ.

Then, $-A$ generates a unique analytic semigroup $\{\exp(-tA)\}$ with the ω of Definition 2.5.1 being the ω of Theorem 2.5.1. In addition, for any $x \in X$ and $0 < \varepsilon < \omega$, we have $\exp(-tA)x \to x$ as $t \to 0$ with $t \in S_{\omega-\varepsilon}$.

Proof: We define $\exp(-tA)$ by the Dunford integral (see Appendix X)

$$\exp(-tA) = (2\pi i)^{-1} \int_C \exp(\lambda t)(\lambda I + A)^{-1}\, d\lambda \qquad (2.5.4)$$

where C is a contour, running in $\rho(-A)$ and consisting of the ray $\{z : \arg z = \theta, \ |z| \leqslant 1\}$ traversed inward from ∞ to $\exp(i\theta)$ for some $\theta \in (\pi/2, (\pi/2)+\omega)$, the circular arc from $\exp(i\theta)$ to $\exp(-i\theta)$ traversed in the positive sense and the ray $\{z : \arg z = -\theta, \ |z| \geqslant 1\}$ traversed outward from $\exp(-i\theta)$ to ∞. In view of (2.5.3) the integral in (2.5.4) converges absolutely and uniformly in any bounded closed subset of the sector $S_{\theta-(\pi/2,)}$ and is consequently analytic in S. The semigroup property follows from the standard argument with Dunford integrals. In fact, if C' denotes any contour obtained from C by a slight shift to the right then

$$\exp(-sA) = (2\pi i)^{-1} \int_{C'} \exp(\lambda's)(\lambda' I + A)^{-1}\, d\lambda', \qquad |\arg s| \leqslant \omega - \varepsilon.$$

$$(2.5.5)$$

Multiplying (2.5.4) and (2.5.5) and using Fubini's theorem and the resolvent formula (see Appendix VIII) we get

$\exp(-tA)\exp(-sA)$

$$= (2\pi i)^{-2} \int_C \int_{C'} \exp(\lambda t + \lambda' s)(\lambda I + A)^{-1}(\lambda' I + A)^{-1} \, d\lambda \, d\lambda'$$

$$= (2\pi i)^{-2} \int_C \int_{C'} \exp(\lambda t + \lambda' s)(\lambda' - \lambda)^{-1}[(\lambda I + A)^{-1} - (\lambda' I + A)^{-1}] \, d\lambda \, d\lambda'.$$

Since C lies to the left of C',

$$\int_C \exp(\lambda t)/(\lambda' - \lambda) \, d\lambda = 0,$$

$$\int_{C'} \exp(\lambda' s)/(\lambda' - \lambda) \, d\lambda' = 2\pi i \exp(\lambda s).$$

Hence, using Fubini's theorem

$$\exp(-tA)\exp(-sA) = (2\pi i)^{-1} \int_C \exp[\lambda(t+s)](\lambda I + A)^{-1} \, d\lambda$$

$$= \exp[-(t+s)A],$$

proving the semigroup property for $\exp(-tA)$. From (2.5.4) it follows that $\exp(-tA)$ is differentiable for $t = 0$ and

$$(d/dt)\exp(-tA) = (2\pi i)^{-1} \int_C \lambda \exp(\lambda t)(\lambda I + A)^{-1} \, d\lambda \in B(X).$$

From the identities $\lambda(\lambda I + A)^{-1} = I - A(\lambda I + A)^{-1}$ and $\int_C \exp(\lambda t) \, d\lambda = 0$, and the closeness of A, we obtain

$$(d/dt)[\exp(-tA)x] = -(2\pi i)^{-1} \int_C A \exp(\lambda t)(\lambda I + A)^{-1} x \, d\lambda$$

$$= -A \exp(-tA)x, \qquad x \in X.$$

To prove the uniform boundedness of $\exp(-tA)$ and $tA \exp(-tA)$ we change the integration variable from λ to $\lambda' = \lambda|t|$ in (2.5.4) and denote by C' the new contour, that is, $C' = |t|C$. In view of Cauchy's theorem, the contour C' can be deformed into the contour C. Thus

$$\exp(-tA) = (2\pi i)^{-1} \int_C \exp[\lambda'(t/|t|)][(\lambda'/|t|)I + A]^{-1}|t|^{-1} \, d\lambda',$$

$$t \in S_{\omega - \varepsilon}. \tag{2.5.6}$$

From (2.5.3)

$$\| [(\lambda'/|t|)I+A]^{-1} \| \leqslant M|t|/|\lambda'|$$

and so

$$\|\exp(-tA)\| \leqslant (M/2\pi) \int_C |\exp[\lambda'(t/|t|)]| \ |\lambda'|^{-1} \ |d\lambda'| = K(\varepsilon).$$

Also, from (2.5.6) and the closeness of A we obtain

$$
\begin{aligned}
A\exp(-tA) &= (2\pi i)^{-1} \int_C \exp[\lambda'(t/|t|)] \, A \, [(\lambda'/|t|)I+A]^{-1}|t|^{-1} \, d\lambda' \\
&= (2\pi i)^{-1} \int_C \exp[\lambda'(t/|t|)] \, ([(\lambda'/|t|)I+A] - (\lambda'/|t|)I) \\
&\quad \times [(\lambda'/|t|)I+A]^{-1}|t|^{-1} \, d\lambda \\
&= (2\pi i)^{-1} \int_C \exp[\lambda'(t/|t|)] \, (I - (\lambda'/|t|)[(\lambda'/|t|)I+A]^{-1}) \\
&\quad \times |t|^{-1} \, d\lambda.
\end{aligned}
$$

Hence

$$\|A\exp(-tA)\| \leqslant K'(\varepsilon)/|t|.$$

To prove that $\exp(-tA)x \to x$ as $t \to 0$ for $t \in S_{\omega-\varepsilon}$ we note that, in view of Cauchy's theorem and Cauchy's integral formula, for $v \in D(A)$

$$
\begin{aligned}
[\exp(-tA) - I]v &= (2\pi i)^{-1} \int_C \exp(\lambda t)[(\lambda I+A)^{-1} - \lambda^{-1}]v \, d\lambda \\
&= -(2\pi i)^{-1} \int_C \exp(\lambda t)(\lambda I+A)^{-1}Av \, \lambda^{-1} \, d\lambda \\
&\to -(2\pi i)^{-1} \int_C (\lambda I+A)^{-1}Av \, \lambda^{-1} \, d\lambda = 0.
\end{aligned}
$$

Since $\exp(-tA)$ is bounded and the domain of A is dense in X, it follows that

$$\lim_{t \to 0} \exp(-tA)x = x, \qquad t \in S_{\omega-\varepsilon}.$$

Strong continuity at any t then results from this and the semigroup property $\exp(-tA)\exp(-sA) = \exp[-(t+s)A]$. Next we prove that $-A$ is the infinitesimal generator of $\{\exp(-tA)\}$. Indeed, for $x \in D(A)$

$$[\exp(-tA)-I]/t\,x = -t^{-1}\int_0^t \exp(-sA)x\,ds \to -Ax.$$

Hence the infinitesimal generator B of $\{\exp(-tA)\}$ satisfies $D(B) \supset D(A)$ and $B = -A$ on $D(A)$. The proof that $D(B) = D(A)$ follows as in Claim 8 of Theorem 2.3.1. Finally, the uniqueness of $\exp(-tA)$ follows as in Theorem 2.2.4.

The proof is therefore complete.

PROBLEM 2.5.1. If the hypothesis (iii) of Theorem 2.5.1 is replaced by the stronger condition

$$\|(\lambda I + A)^{-1}\| \leqslant M/(|\lambda|+p), \qquad \lambda \in S_\omega \tag{2.5.7}$$

for some $p > 0$, then instead of (2.5.2) one can derive the stronger estimates
$\|\exp(-tA)\| \leqslant K\exp(-\delta\,\mathrm{Re}\,t)$, $\|A\exp(-tA)\| \leqslant (K/|t|)\exp(-\delta\,\mathrm{Re}\,t)$
where δ is some positive number.

As an application of Theorem 2.5.1, we shall consider the abstract Cauchy problem

$$du/dt + Au = f(t), \qquad 0 \leqslant t \leqslant T; \tag{2.5.8}$$

$$u(0) = u_0, \qquad u_0 \in D(A) \tag{2.5.9}$$

where the operator A generates an analytic semigroup. In Theorem 2.2.3 we assumed that $f(t)$ was strongly continuously differentiable but here, where A generates an analytic semigroup, we can relax this hypothesis and assume that $f(t)$ is Hölder continuous.

THEOREM 2.5.2. Assume that $-A$ generates an analytic semigroup and $f(t)$ is uniformly Hölder continuous on $[0, T]$. Then the abstract Cauchy problem (2.5.8) and (2.5.9) has the unique solution

$$u(t) = \exp(-tA)u_0 + \int_0^t \exp[-(t-s)A]f(s)\,ds. \tag{2.5.10}$$

If in addition, $f(t)$ is analytic in a domain containing $(0, T]$, the solution $u(t)$ is analytic at any $t \in (0, T]$.

Proof: It is easy to show (as in Theorem 2.2.2) that $\exp(-tA)u_0$ is the unique solution of the homogeneous problem

$$du/dt + Au = 0 \qquad \text{and} \qquad u(0) = u_0.$$

Therefore, it suffices to prove that the function

$$w(t) = \int_0^t \exp[-(t-s)A]f(s)\,ds$$

satisfies (2.5.9) and (2.5.10) with $u_0 = 0$. We first write

$$w(t) = \int_0^t \exp[-(t-s)A][f(s)-f(t)]\,ds + \int_0^t \exp[-(t-s)A]f(t)\,dt.$$

Then, using the identity

$$A\int_0^t \exp(-sA)x\,ds = x - \exp(-tA)x, \qquad x \in X,$$

we can formally write

$$Aw(t) = \int_0^t A\exp[-(t-s)A][f(s)-f(t)]\,ds + f(t) - \exp(-tA)f(t).$$
$$(2.5.11)$$

In view of the inequalities $\|A\exp[-(t-s)A]\| \leqslant \mathrm{const}(t-s)^{-1}$ and $\|f(s)-f(t)\| \leqslant \mathrm{const}(t-s)^{\beta}$ for $0 < \beta \leqslant 1$, the integral in (2.5.11) is well defined. Thus $Aw(t)$ makes sense, and because A is closed we get

$$Aw(t) = A\int_0^t \exp[-(t-s)A][f(s)-f(t)]\,ds + f(t) - \exp(-tA)f(t).$$

In order to construct $dw(t)/dt$ we define the function

$$w_\varepsilon(t) = \int_0^{t-\varepsilon} \exp[-(t-s)A]f(s)\,ds, \qquad \varepsilon > 0 \text{ and small.} \quad (2.5.12)$$

Clearly $w_\varepsilon(t) \to w(t)$ as $\varepsilon \to 0_+$, uniformly on compact sets in $(0, T]$. Also

$$dw_\varepsilon(t)/dt = \exp(-\varepsilon A)f(t-\varepsilon) - \int_0^{t-\varepsilon} A\exp[-(t-s)A]f(s)\,ds.$$
$$(2.5.13)$$

The integral in (2.5.13) exists since the operator $A\exp[-(t-s)A]$ is bounded for $0 \leqslant s \leqslant t-\varepsilon$. Employing again the Hölder continuity of $f(s)$ as in (2.5.11), it is easy to verify that the integral in (2.5.13) converges to $Aw(t)$ uniformly on compact intervals on $(0, T]$. We have proved that

$$\lim_{\varepsilon \to 0} w_\varepsilon(t) = w(t) \qquad \text{and} \qquad \lim_{\varepsilon \to 0} dw_\varepsilon(t)/dt = f(t) - Aw(t)$$

uniformly on compact intervals of $(0, T]$. From a well-known theorem of calculus it follows that $dw(t)/dt$ exists and

$$dw/dt = f(t) - Aw(t),$$

proving the first part of the theorem. Now if $f(t)$ is analytic in some domain containing $(0, T]$, the function $w_\varepsilon(t)$ in (2.5.12) is also analytic in some neighborhood N of $[\varepsilon, T]$ and $dw_\varepsilon(t)/dt$ exists for $t \in N$. Since $w_\varepsilon(t) \to w(t)$ uniformly on compact sets in N, it follows that $w(t)$ is analytic for any $t \in (0, T]$.

The proof is complete.

2.6. Notes

Most of the results of this chapter concerning semigroups of operators were obtained by E. Hille, K. Yosida and R. S. Phillips. The bible in semigroup theory is still Hille and Phillips [28] where the reader is referred to for more details on the subject. Lemma 2.3.1, Theorem 2.3.2, and Corollary 2.3.1 are due to Kato [30]. The results of Section 2.4 are due to Hale [27] while Section 2.5 consists of the work by Yosida [78] and Hille and Phillips [28]. See also Solomiak [66], Kato [33], Phillips [60], and Dunford and Schwartz [18, 19]. Several recent references and good examples will be found in Friedman [23], Krein [35], Carroll [12], and the lecture notes by Zaidman [79] and Goldstein [26].

Chapter 3

Linear Evolution Equations of the Parabolic Type: Sobolevski–Tanabe Theory

---

3.0. Introduction

A detailed study of the time-dependent Cauchy problem

$$du/dt + A(t)u = f(t), \qquad 0 < t \leqslant T \quad \text{and} \quad u(0) = u_0 \in X \qquad (3.0.1)$$

forms the major content of this chapter. Here, for each t, the operator $-A(t)$ generates an analytic semigroup. Since parabolic partial differential equations can be realized in this form, (3.0.1) is said to be parabolic. We prove the existence and uniqueness of a fundamental solution of the evolution equation (the terminology after L. Schwarz)

$$du/dt + A(t)u = 0, \qquad 0 < t \leqslant T$$

as well as existence, uniqueness and differentiability of solutions of (3.0.1). Section 3.8 deals with the asymptotic behavior of solutions of (3.0.1) when $T = \infty$.

3.1. Definitions and Hypotheses

Consider the *evolution equation*

$$du/dt + A(t)u = f(t), \qquad 0 < t \leqslant T \qquad (3.1.1)$$

and the associated homogeneous equation

$$du/dt + A(t)u = 0, \qquad 0 < t \leqslant T \qquad (3.1.2)$$

where the unknown $u(t)$ is a function $u\colon [0, T] \to X$, from the real interval $[0, T]$ into a Banach space X. $f\colon [0, T] \to X$ is a given function and for each t such that $0 \leqslant t \leqslant T$, $A(t)$ is a given, closed, linear operator in X with domain $D[A(t)] = D$, independent of t and dense in X. The problem of finding a solution u of the initial value problem

$$du/dt + A(t)u = f(t), \qquad 0 < t \leqslant T; \qquad (3.1.3)$$

$$u(0) = u_0, \qquad u_0 \in X \qquad (3.1.4)$$

is called an *abstract Cauchy problem*.

DEFINITION 3.1.1. An operator-valued function $U(t, \tau)$ with values in $B(X)$, defined and strongly continuous jointly in t, τ for $0 \leqslant \tau \leqslant t \leqslant T$, is called a *fundamental solution* of (3.1.2) if

(i)   the partial derivative $\partial U(t, \tau)/\partial t$ exists in the strong topology of X, belongs to $B(X)$ for $0 \leqslant \tau < t \leqslant T$, and is strongly continuous in t for $0 \leqslant \tau < t \leqslant T$;
(ii)   the range of $U(t, \tau)$ is in D;

(iii) $\partial U(t, \tau)/\partial t + A(t) U(t, \tau) = 0, \qquad 0 \leqslant \tau < t \leqslant T \qquad (3.1.5)$

and

$$U(\tau, \tau) = I. \qquad (3.1.6)$$

$U(t, \tau)$ is also called evolution operator, propagator, solution operator, Green's function, etc.

DEFINITION 3.1.2.   A function $u: [0, T] \to X$ is called a *(strict) solution* of the abstract Cauchy problem (3.1.3) and (3.1.4) if $u(t)$ is strongly continuous on $[0, T]$, strongly continuously differentiable on $(0, T]$, $u(t) \in D$ for $0 < t \leqslant T$ and $u(t)$ satisfies the system (3.1.3) and (3.1.4).

DEFINITION 3.1.3.   A function $u: [0, T] \to X$ is called a *mild solution* of (3.1.3) and (3.1.4) if it admits the integral representation

$$u(t) = U(t, 0) u_0 + \int_0^t U(t, s) f(s) \, ds. \tag{3.1.7}$$

It should be remarked that (3.1.7) need not give a solution of (3.1.3) and (3.1.4) for every u_0 and $f(t)$. The existence of du/dt and $A(t)u$ for (3.1.7) can be proved only under certain assumptions on u_0 and $f(t)$.

There are various known sufficient conditions for the existence of the evolution operator $U(t, \tau)$. In practically all cases so far considered in the literature $-A(t)$ is assumed to be the infinitesimal generator of a strongly continuous semigroup of bounded linear operators on X. In addition $A(t)$ is assumed to depend on t smoothly in some sense. Roughly speaking, there are two important cases to be distinguished: the hyperbolic and the parabolic. In the *hyperbolic* case $-A(t)$ is, for each t, the infinitesimal generator of a contraction semigroup. In the *parabolic* case $-A(t)$ is, for each t, the infinitesimal generator of an analytic semigroup. Actually, the two cases are not mutually exclusive, and in many instances "parabolic" is a stronger notion than "hyperbolic."

In this chapter we shall study the parabolic case and we shall refer the reader to the fundamental paper of Kato [30] for the hyperbolic case. By S we denote the set of all complex numbers λ satisfying $-\theta \leqslant \arg \lambda \leqslant \theta$, where θ is a fixed angle with $\pi/2 < \theta < \pi$. Throughout this chapter we shall make constant use of the following hypotheses:

Hypothesis 1:   For each $\sigma \in [0, T]$ we have $A(\sigma)$ a closed operator in X with domain $D[A(\sigma)] = D$ independent of σ and dense in X.

Hypothesis 2:   For each $\sigma \in [0, T]$, the resolvent set $\rho[-A(\sigma)]$ of $-A(\sigma)$ contains S and

$$\|[\lambda I + A(\sigma)]^{-1}\| \leqslant C/(1 + |\lambda|), \qquad \lambda \in S,$$

where C is a positive constant independent of λ and σ.

Hypothesis 3: The bounded operator $A(t)A^{-1}(s)$ is Hölder continuous in t in the uniform operator topology for each fixed s, that is,

$$\|[A(t)-A(\tau)]A^{-1}(s)\| \leqslant C|t-\tau|^{\alpha}, \qquad 0 < \alpha \leqslant 1,$$

where C and α are positive constants independent of t, τ, and s for $0 \leqslant t, \tau, s \leqslant T$.

Hypothesis 4: The function $f(t)$ is uniformly Hölder continuous on $[0, T]$:

$$\|f(t)-f(s)\| \leqslant C|t-s|^{\beta}, \qquad 0 < \beta \leqslant 1, \quad 0 \leqslant t, s \leqslant T,$$

where C and β are positive constants independent of t and s.

Hypothesis 5: The function $f(t)$ is k-times (strongly) continuously differentiable on $[0, T]$, and $f^{(k)}(t)$ is uniformly Hölder continuous on $[0, T]$, that is,

$$\|f^{(k)}(t)-f^{(k)}(s)\| \leqslant C|t-s|^{\beta}, \qquad 0 < \beta \leqslant 1, \quad 0 \leqslant t, s \leqslant T,$$

where C and β are positive constants independent of t and s.

Hypothesis 6: The operator $A(t)$ with $0 \leqslant t \leqslant T$ is k-smooth in the following sense: for each $x \in X$ the function $A(t)A^{-1}(0)x$ has strongly continuous derivatives

$$(d^{j}/dt^{j})[A(t)A^{-1}(0)x] \equiv A^{(j)}(t)A^{-1}(0)x, \qquad 1 \leqslant j \leqslant k.$$

The operators $A^{(j)}(t)A^{-1}(0)x$ with $1 \leqslant j \leqslant k$, are uniformly bounded for $0 \leqslant t \leqslant T$, and

$$\|A^{(k)}(t)A^{-1}(0) - A^{k}(\tau)A^{-1}(0)\| \leqslant C|t-\tau|^{\alpha}, \qquad 0 < \alpha \leqslant 1,$$

where C and α are positive constants independent of t and τ for $0 \leqslant t, \tau \leqslant T$.

To study the asymptotic behavior of the solutions of the evolution equation

$$du/dt + A(t)u = 0, \qquad 0 < t < \infty \tag{3.1.8}$$

we shall need the following hypotheses:

Hypothesis 7: Hypotheses 1–3 are satisfied for all $0 < T < \infty$ with C and α positive constants independent of T. Furthermore

$$\sup_{0 < t,\, \tau < \infty} \|A(t)A^{-1}(\tau)\| < \infty. \tag{3.1.9}$$

Hypothesis 8: Hypothesis 4 is satisfied for $0 \leqslant t, s < \infty$. Furthermore, there exists an element $f(\infty) \in X$ such that

$$\lim_{t \to \infty} \| f(t) - f(\infty) \| = 0. \tag{3.1.10}$$

Hypothesis 9: There exists a closed operator $A(\infty)$ with domain D and with bounded inverse such that

$$\lim_{t \to \infty} \| [A(t) - A(\infty)] A^{-1}(0) \| = 0. \tag{3.1.11}$$

REMARK 3.1.1. Hypotheses 1 and 2 imply that, for each $\sigma \in [0, T]$, $-A(\sigma)$ generates an analytic semigroup $\{\exp[-tA(\sigma)]\}$, $t \geqslant 0$. From the results of Section 2.5 we single out the following consequences which will be often used in this chapter. There exist positive numbers δ and C independent of t and σ such that for each $\sigma \in [0, T]$

$$(d/dt)\exp[-tA(\sigma)]x = -A(\sigma)\exp[-tA(\sigma)]x, \qquad x \in X, \quad t > 0; \tag{3.1.12}$$

$$\|\exp[-tA(\sigma)]\| \leqslant C \exp(-\delta t), \qquad t > 0; \tag{3.1.13}$$

$$\|A(\sigma)\exp[-tA(\sigma)]\| \leqslant C t^{-1} \exp(-\delta t), \qquad t > 0. \tag{3.1.14}$$

In the remainder of this chapter, C will denote a generic constant, that is, a constant independent of $t, \tau, \sigma, \ldots$.

REMARK 3.1.2. From Hypothesis 3, it follows, by taking $s = \tau$ and using the triangle inequality, that

$$\|A(t)A^{-1}(\tau)\| \leqslant C. \tag{3.1.15}$$

Writing

$$A(\tau)\exp[-tA(\sigma)] = A(\tau)A^{-1}(\sigma) \cdot A(\sigma)\exp[-tA(\sigma)]$$

and using (3.1.15) and (3.1.14), we obtain

$$\|A(\tau)\exp[-tA(\sigma)]\| \leqslant C t^{-1} \exp(-\delta t), \qquad t > 0. \tag{3.1.16}$$

For any positive integer n we may write

$$A^n(\sigma)\exp[-tA(\sigma)]x = A(\sigma)\exp[-tn^{-1}A(\sigma)] \cdots A(\sigma)\exp[-tn^{-1}A(\sigma)]x$$

$$(n \text{ factor}), \qquad x \in X.$$

In fact, for $x \in X$ $\exp[-tn^{-1}A(\sigma)]x \in D$ and $A(\sigma)$ commute with $\exp[-tn^{-1}A(\sigma)]$ on D. We conclude that $\exp[-tA(\sigma)]x \in D(A^n)$ for $t > 0$ and from (3.1.14) we obtain the useful estimate

$$\|A^n(\sigma)\exp[-tA(\sigma)]\| \leqslant Ct^{-n}\exp(-\delta t), \qquad t > 0. \qquad (3.1.17)$$

REMARK 3.1.3.   Hypothesis 3 is equivalent to

Hypothesis 3′:

$$\|[A(t)-A(\tau)]A^{-1}(\tau)\| \leqslant C|t-\tau|^{\alpha}, \qquad 0 < \alpha \leqslant 1, \quad 0 \leqslant t,\tau \leqslant T,$$

where C and α are positive constants independent of t and τ. In fact, Hypothesis 3 implies Hypothesis 3′ and if Hypothesis 3′ is valid, we have, using 3.1.15,

$$\begin{aligned}
\|[A(t)-A(\tau)]A^{-1}(s)\| &= \|[A(t)-A(\tau)]A^{-1}(\tau) \cdot A(\tau)A^{-1}(s)\| \\
&\leqslant C|t-\tau|^{\alpha} \cdot C \\
&= C|t-\tau|^{\alpha}
\end{aligned}$$

and our assertion is established.

3.2. Statements of the Main Theorems and Some Heuristic Arguments

Here we shall state the main results that we plan to prove in this chapter. We also present some heuristic arguments which will be substantiated in the following sections under the strength of the given hypotheses.

THEOREM 3.2.1.   Let Hypotheses 1–3 be satisfied. Then the evolution equation (3.1.2) has a unique fundamental solution $U(t,\tau)$.

THEOREM 3.2.2.   Let Hypotheses 1–4 be satisfied. Then the abstract Cauchy problem (3.1.3) and (3.1.4) has the unique solution

$$u(t) = U(t,0)u_0 + \int_0^t U(t,s)f(s)\, ds. \qquad (3.2.1)$$

THEOREM 3.2.3.   Let Hypotheses 1–3, 5, and 6 be satisfied. Then the solution $u(t)$ of the abstract Cauchy problem (3.1.3) and (3.1.4) is $(k+1)$-times (strongly) continuously differentiable on $[\varepsilon, T]$ for any $\varepsilon > 0$.

THEOREM 3.2.4.   Let Hypotheses 7–9 be satisfied. Then any solution $u(t)$ of (3.1.8) converges to some element $u(\infty) \in D$ such that

$$A(\infty)u(\infty) = f(\infty). \qquad (3.2.2)$$

Moreover, $du(t)/dt$ tends to 0 as $t \to \infty$.

The following arguments, although without mathematical rigor, will give us the guidelines which we should follow to prove the existence and uniqueness Theorems 3.2.1 and 3.2.2. Since $U(t, \tau)$ is the fundamental solution of (3.1.2) and the function $\exp[-(t-\tau)A(\tau)]$ satisfies the operator system

$$dv/dt + A(\tau)v = 0 \qquad \text{and} \qquad v(\tau) = I,$$

it follows that the function $v(t) = U(t, \tau) - \exp[-(t-\tau)A(\tau)]$ must satisfy the system

$$dv/dt + A(t)v = [A(\tau) - A(t)]\exp[-(t-\tau)A(\tau)] \qquad \text{and} \qquad v(\tau) = 0. \qquad (3.2.3)$$

But, from the variation of constants formula, the solution $v(t)$ of system (3.2.3) is given by

$$v(t) = \int_{\tau}^{t} U(t, s)[A(\tau) - A(s)]\exp[-(s-\tau)A(\tau)]\,ds.$$

Hence

$$U(t, \tau) = \exp[-(t-\tau)A(\tau)] + \int_{\tau}^{t} U(t, s)[A(\tau) - A(s)]$$
$$\times \exp[-(s-\tau)A(\tau)]\,ds. \qquad (3.2.4)$$

We shall consider (3.2.4) as the defining relation for the unknown operator $U(t, \tau)$. This is an integral operator equation of the Volterra type (with respect to τ). Applying the method of successive approximations the solution of (3.2.4) is formally given by the series

$$U(t, \tau) = \sum_{k=0}^{\infty} U_k(t, \tau) \qquad (3.2.5)$$

where

$$U_0(t, \tau) = \exp[-(t-\tau)A(\tau)]$$

and

$$U_k(t, \tau) = \int_{\tau}^{t} U_{k-1}(t, s)[A(\tau) - A(s)]\exp[-(s-\tau)A(\tau)]\,ds. \qquad (3.2.6)$$

Set

$$\phi_1(t, \tau) = [A(\tau) - A(t)] \exp[-(t - \tau) A(\tau)], \qquad (3.2.7)$$

$$\phi_{k+1}(t, \tau) = \int_\tau^t \phi_k(t, s) \phi_1(s, \tau) \, ds, \qquad k = 1, 2, \dots, \qquad (3.2.8)$$

and

$$\Phi(t, \tau) = \sum_{k=1}^\infty \phi_k(t, \tau). \qquad (3.2.9)$$

Hence, formally

$$\Phi(t, \tau) = \phi_1(t, \tau) + \int_\tau^t \Phi(t, s) \phi_1(s, \tau) \, ds. \qquad (3.2.10)$$

On the other hand, using Fubini's theorem, one can show by induction (provided that all make sense) that

$$U_k(t, \tau) = \int_\tau^t \exp[-(t - s) A(s)] \phi_k(s, \tau) \, ds, \qquad k = 1, 2, \dots .$$

Hence, formally

$$U(t, \tau) = \exp[-(t - \tau) A(t)] + \int_\tau^t \exp[-(t - s) A(s)] \Phi(s, \tau) \, ds. \quad (3.2.11)$$

Our program is now clear. First, we will prove that the Volterra integral equation (3.2.10) has a solution $\Phi(t, \tau)$. Second, the formula (3.2.11) gives the desired fundamental solution of the evolution equation (3.1.2). Third, using the Hölder continuity of $f(t)$ we shall prove that the integral in (3.2.1) makes sense and (3.2.1) defines the unique solution of (3.1.3) and (3.1.4). The proofs of the other theorems do not involve many complications.

3.3. Properties of the Semigroup $\{\exp[-tA(\tau)]\}$

In this section we shall establish a series of interesting lemmas and estimates concerning the semigroup $\{\exp[-tA(\tau)]\}$ which will be useful in the subsequent sections. Here we assume that Hypotheses 1–3 are satisfied. The variables $t, s, \tau, \xi, \eta, \dots$ are assumed in $[0, T]$.

LEMMA 3.3.1.   For $v \in D$, the following identities hold:

(a) $$(\exp[-\tau A(t)] - \exp[-\tau A(s)])v$$

$$= \int_0^\tau \exp[-(\tau-\xi)A(t)][A(s)-A(t)]A^{-1}(s)$$

$$\times \exp[-\xi A(s)]A(s)v \, d\xi. \tag{3.3.1}$$

(b) $$(\exp[-tA(\tau)] - \exp[-sA(\tau)])v$$

$$= -\int_s^t \exp[-\xi A(\tau)]A(\tau)v \, d\xi. \tag{3.3.2}$$

Proof: (a) Define the function

$$F(\xi) = -\exp[-(\tau-\xi)A(t)]\exp[-\xi A(s)]v, \quad \tau \geqslant \xi.$$

F is strongly differentiable with respect to ξ with derivative

$$F'(\xi) = -\exp[-(\tau-\xi)A(t)]A(t)\exp[-\xi A(s)]v$$
$$+ \exp[-(\tau-\xi)A(t)]A(s)\exp[-\xi A(s)]v$$
$$= \exp[-(\tau-\xi)A(t)][A(s)-A(t)]\exp[-\xi A(s)]v$$
$$= \exp[-(\tau-\xi)A(t)][A(s)-A(t)]A^{-1}(s)$$
$$\times \exp[-\xi A(s)]A(s)v. \tag{3.3.3}$$

Since $F'(\xi)$ is continuous, (3.3.1) follows upon integrating (3.3.3) from 0 to τ and using Theorem 1.3.4.

(b) Define the function

$$\Phi(\xi) = \exp[-\xi A(\tau)]v.$$

Then

$$\Phi'(\xi) = -\exp[-\xi A(\tau)]A(\tau)v \tag{3.3.4}$$

and $\Phi'(\xi)$ is continuous in ξ. Integrating (3.3.4) from s to t, (3.3.2) follows.

The identities (3.3.1) and (3.3.2) permit us to establish the following:

LEMMA 3.3.2. The following inequalities hold:

(a) $$\|\exp[-\tau A(t)] - \exp[-\tau A(s)]\| \leqslant C|t-s|^\alpha \exp(-\delta t); \tag{3.3.5}$$

(b) $$\|A(\xi)(\exp[-\tau A(t)] - \exp[\tau A(s)])\|$$
$$\leqslant C\tau^{-1}|t-s|^\alpha \exp(-\delta\tau), \quad \tau > 0; \tag{3.3.6}$$

(c) $$\|A(\xi)(\exp[-\tau A(t)] - \exp[-\tau A(s)])A^{-1}(\eta)\|$$
$$\leqslant C|t-s|^\alpha \exp(-\delta\tau). \tag{3.3.7}$$

(d) $\|(\exp[-\tau A(t)] - \exp[-sA(\tau)]) A^{-1}(\eta)\|$

$\qquad \leqslant C|t-s| \exp[-\delta \min(t,s)];$ $\qquad\qquad$ (3.3.8)

(e) $\|A(\eta)(\exp[-tA(\tau)] - \exp[-sA(\tau)]) A^{-2}(\tau)\|$

$\qquad \leqslant C|t-s| \exp[-\delta \min(t,s)].$ $\qquad\qquad$ (3.3.9)

Proof: (a)   Let $x \in X$. Then

$$(\exp[-\tau A(t)] - \exp[-\tau A(s)]) x$$

$$= (\exp[-(\tau/2) A(t)] - \exp[-(\tau/2) A(s)]) \exp[-(\tau/2) A(s)] x$$

$$+ \exp[-(\tau/2) A(t)] (\exp[-(\tau/2) A(t)] - \exp[-(\tau/2) A(s)]) x$$

$$= (\exp[-(\tau/2) A(t)] - \exp[-(\tau/2) A(s)]) \exp[-(\tau/2) A(s)] x$$

$$+ A(t) \exp[-(\tau/2) A(t)]$$

$$\times (\exp[-(\tau/2) A(t)] - \exp[-(\tau/2) A(s)]) A^{-1}(s) x$$

$$+ \exp[-(\tau/2) A(t)][A(t) - A(s)] A^{-1}(s) \exp[-(\tau/2) A(s)] x$$

$$- \exp[-\tau A(t)][A(t) - A(s)] A^{-1}(s) x$$

$$\equiv I_1 + I_2 + I_3. \qquad\qquad (3.3.10)$$

Using (3.3.1) with $v = \exp[-(\tau/2) A(s)]$ for $x \in D$ we obtain

$$I_1 = \int_0^{\tau/2} \exp[-(\tau/2 - \xi) A(t)][A(s) - A(t)] A^{-1}(s) \exp[-\xi A(s)]$$

$$\times A(s) \exp[-(\tau/2) A(s)] x \, d\xi.$$

In view of (3.1.13), Hypothesis 3, and (3.1.14) we get

$$\|I_1\| \leqslant \int_0^{\tau/2} C \exp[-\delta(\tau/2 - \xi)] C|s - t|^\alpha \exp(-\delta\xi)$$

$$\times C(\tau/2)^{-1} \exp(-\delta\tau/2) \|x\| \, d\xi$$

$$\leqslant C|s - t|^\alpha \exp(-\delta\tau) \|x\|. \qquad\qquad (3.3.11)$$

Employing (3.3.1) with $v = A^{-1}(s) x \in D$ we see that

$$I_2 = A(t) \exp[-(\tau/2) A(t)] \int_0^{\tau/2} \exp[-(\tau/2 - \xi) A(t)]$$

$$\times [A(s) - A(t)] A^{-1}(s) \exp[-\xi A(s)] x \, d\xi.$$

Because of (3.1.14), (3.1.13), and Hypothesis 3 it follows that

$$\|I_2\| \leqslant C\exp(-\delta\tau/2)(2/\tau)\int_0^{\tau/2} C\exp[-\delta(\tau/2-\xi)]$$

$$\times C|s-t|^\alpha \exp(-\delta\xi)\|x\|\,d\xi$$

$$\leqslant C|s-t|^\alpha \exp(-\delta\tau)\|x\|. \tag{3.3.12}$$

Finally, in view of (3.1.13) and Hypothesis 3 we have

$$\|I_3\| \leqslant C\exp(-\delta\tau/2)\,C|t-s|^\alpha C\exp(-\delta\tau/2)\|x\|$$

$$+ C\exp(-\delta\tau)\,C|t-s|^\alpha\|x\|$$

$$= C|s-t|^\alpha \exp(-\delta\tau)\|x\|. \tag{3.3.13}$$

From (3.3.10)–(3.3.13), we conclude that for every $x \in X$

$$\|(\exp[-\tau A(t)] - \exp[-\tau A(s)])x\| \leqslant \|I_1\| + \|I_2\| + \|I_3\|$$

$$\leqslant C|t-s|^\alpha \exp(-\delta\tau)\|x\|,$$

establishing the desired bound (3.3.5).

(b) Using the notation in (3.3.10), we notice that, for every $x \in X$

$$A(\xi)(\exp[-\tau A(t)] - \exp[-\tau A(s)])x$$

$$= A(\xi)I_1 + A(\xi)I_2 + A(\xi)I_3. \tag{3.3.14}$$

Now

$$A(\xi)I_2 = A(\xi)A^{-1}(t)A^2(t)\exp[-(\tau/2)A(t)]$$

$$\times \int_0^{\tau/2} \exp[-(\tau/2-\xi)A(t)][A(s)-A(t)]A^{-1}(s)$$

$$\times \exp[-\xi A(s)]x\,d\xi$$

and on account of (3.1.15), (3.1.17) with $n = 2$, (3.1.13) and Hypothesis 3, we get

$$\|A(\xi)I_2\| \leqslant C(4C/\tau^2)\exp(-\delta\tau/2)\int_0^{\tau/2} C\exp[-\delta(\tau/2-\xi)]$$

$$\times C|s-t|^\alpha \exp(-\delta\xi)\|x\|\,d\xi$$

$$= (C/\tau^2)\exp(-\delta\tau)|s-t|^\alpha\|x\|\,\tau/2$$

$$= C/\tau\,|t-s|^\alpha \exp(-\delta\tau)\|x\|. \tag{3.3.15}$$

Also

$$A(\xi)I_3 = A(\xi)\exp[-(\tau/2)A(t)][A(t)-A(s)]A^{-1}(s)\exp[-(\tau/2)A(s)]x$$
$$- A(\xi)\exp[-\tau A(t)][A(t)-A(s)]A^{-1}(s)x;$$

therefore

$$\|A(\xi)I_3\| \leqslant (2C/\tau)\exp(-\delta\tau/2)\,C|t-s|^{\alpha}C\exp(-\delta\tau/2)\|x\|$$
$$+ (C/\tau)|t-s|^{\alpha}\exp(-\delta\tau)\|x\|$$
$$= (C/\tau)|t-s|^{\alpha}\exp(-\delta\tau)\|x\|. \tag{3.3.16}$$

Finally

$$A(\xi)I_1 = A(\xi)A^{-1}(t)A(t)(\exp[-(\tau/2)A(t)] - \exp[-(\tau/2)A(s)])$$
$$\times \exp[-(\tau/2)A(s)]x$$
$$= A(\xi)A^{-1}(t)I_4 \tag{3.3.17}$$

where

$$I_4 = A(t)(\exp[(-\tau/2)A(t)] - \exp[-(\tau/2)A(s)])\exp[-(\tau/2)A(s)]x$$
$$= (\exp[-(\tau/2)A(t)] - \exp[-(\tau/2)A(s)])A(s)\exp[-(\tau/2)A(s)]x$$
$$+ [A(s)-A(t)]\exp[-\tau A(s)]x$$
$$+ \exp[-(\tau/2)A(t)][A(t)-A(s)]\exp[-(\tau/2)A(s)]x$$
$$\equiv I_5 + I_6 + I_7. \tag{3.3.18}$$

It follows that

$$\|I_6\| = \|[A(s)-A(t)]A^{-1}(s)\cdot A(s)\exp[-\tau A(s)]x\|$$
$$\leqslant C|t-s|^{\alpha}\cdot(C/\tau)\exp(-\delta\tau)\|x\|$$
$$= (C/\tau)|t-s|^{\alpha}\exp(-\delta\tau)\|x\|. \tag{3.3.19}$$

Also

$$\|I_7\| = \|\exp[-(\tau/2)A(t)][A(t)-A(s)]A^{-1}(s)\cdot A(s)\exp[-(\tau/2)A(s)]x\|$$
$$\leqslant C\exp(-\delta\tau/2)\,C|t-s|^{\alpha}(2C/\tau)\exp(-\delta\tau/2)\|x\|$$
$$= (C/\tau)|t-s|^{\alpha}\exp(-\delta\tau)\|x\|. \tag{3.3.20}$$

To estimate I_5 we use (3.3.1) with $v = A(s)\exp[-(\tau/2)A(s)]$ for $x \in D$ (recall that for $x \in X$ we have $\exp[-(\tau/2)A(s)]x \in D(A^2)$). We have

$$I_5 = \int_0^{\tau/2} \exp[-(\tau/2-\xi)A(t)][A(s)-A(t)]A^{-1}(s)\exp[-\xi A(s)]A^2(s)$$
$$\times \exp[-(\tau/2)A(s)]x \, d\xi.$$

Hence

$$\|I_5\| \leqslant \int_0^{\tau/2} C\exp[-\delta(\tau/2-\xi)]\,C|s-t|^\alpha C\exp(-\delta\xi)$$
$$\times (4C/\tau^2)\exp(-\delta\tau/2)\|x\| \, d\xi$$
$$\leqslant (C/\tau)|t-s|^\alpha\exp(-\delta\tau)\|x\|. \tag{3.3.21}$$

Using (3.3.17)–(3.3.21) we obtain

$$\|A(\xi)I_1\| \leqslant C\|I_4\|$$
$$\leqslant C(\|I_5\| + \|I_6\| + \|I_7\|)$$
$$\leqslant (C/\tau)|t-s|^\alpha\exp(-\delta\tau)\|x\|. \tag{3.3.22}$$

In view of (3.3.14), (3.3.22), (3.3.15), and (3.3.16) the desired estimate (3.3.6) follows.

(c) Let $x \in X$. Then

$$A(\xi)(\exp[-\tau A(t)] - \exp[-\tau A(s)])A^{-1}(\eta)x$$
$$= A(\xi)(\exp[-(\tau/2)A(t)] - \exp[-(\tau/2)A(s)])\exp[-(\tau/2)A(s)]$$
$$\times A^{-1}(\eta)x$$
$$+ A(\xi)\exp[-(\tau/2)A(t)](\exp[-(\tau/2)A(t)] - \exp[-(\tau/2)A(s)])$$
$$\times A^{-1}(\eta)x$$
$$\equiv J_1 + J_2. \tag{3.3.23}$$

Treating J_1 in the same way we treated I_4 in (b), we obtain

$$\|J_1\| \leqslant C|t-s|^\alpha\exp(-\delta\tau)\|x\|. \tag{3.3.24}$$

Using (3.3.1) with $v = A^{-1}(\eta)x \in D$, we see that

$$J_2 = A(\xi)\exp[-(\tau/2)A(t)]\int_0^{\tau/2}\exp[-(\tau/2-\xi)A(t)]$$
$$\times [A(s)-A(t)]A^{-1}(s)\exp[-\xi A(s)]A(s)A^{-1}(\eta)x \, d\xi.$$

Hence

$$\|J_2\| \leqslant (C/\tau)\exp(-\delta\tau/2)\int_0^{\tau/2} C\exp[-\delta(\tau/2-\xi)]$$

$$\times C|t-s|^\alpha C\exp(-\delta\xi)C\|x\|\,d\xi$$

$$\leqslant C|t-s|^\alpha\exp(-\delta\tau)\|x\|. \tag{3.3.25}$$

By (3.3.23), (3.3.24), and (3.3.25) the inequality (3.3.7) follows.

(d) Let $x \in X$. Then $v = A^{-1}(\eta)x \in D$ and (3.3.2) becomes

$$(\exp[-tA(\tau)] - \exp[-sA(\tau)])A^{-1}(\eta)x$$

$$= -\int_s^t \exp[-\xi A(\tau)]A(\tau)A^{-1}(\eta)x\,d\xi.$$

Consequently

$$\|(\exp[-tA(\tau)] - \exp[-sA(\tau)])A^{-1}(\eta)x\|$$

$$\leqslant \left|\int_s^t C\exp(-\delta\xi)C\,\|x\|\,d\xi\right|$$

$$\leqslant C|t-s|\exp[-\delta\min(t,s)]\,\|x\|$$

and (3.3.8) has been established.

(e) Let $x \in X$. Then $v = A^{-2}(\tau)x \in D$ and (3.3.2) reduces to

$$A(\eta)(\exp[-tA(\tau)] - \exp[-sA(\tau)])A^{-2}(\tau)x$$

$$= -\int_s^t A(\eta)\exp[-\xi A(\tau)]A(\tau)A^{-2}(\tau)x\,d\xi$$

$$= -\int_s^t A(\eta)\exp[-\xi A(\tau)]A^{-1}(\tau)x\,d\xi$$

$$= -\int_s^t A(\eta)A^{-1}(\tau)\exp[-\xi A(\tau)]x\,d\xi.$$

Hence

$$\|A(\eta)(\exp[-tA(\tau)] - \exp[-sA(\tau)])A^{-2}(\tau)x\|$$

$$\leqslant \left|\int_s^t C[C\exp(-\delta\xi)]\,\|x\|\,d\xi\right|$$

$$\leqslant C|t-s|\exp[-\delta\min(t,s)]\,\|x\|$$

and (3.3.9) has been established.

The proof of Lemma 3.3.2 is complete.

PROBLEM 3.3.1. Show that the operator $A^{-1}(t)$ is uniformly Hölder continuous on $[0, T]$, that is,

$$\|A^{-1}(t) - A^{-1}(\tau)\| \leqslant C|t - \tau|^{\alpha}, \qquad 0 \leqslant t, \tau \leqslant T.$$

[Hint: Use (2.3.7) with $\lambda = 0$ and (3.3.5).]

LEMMA 3.3.3. For $\tau \geqslant \varepsilon > 0$, the operator-valued function $A(t)\exp[-\tau A(s)]$ is uniformly continuous in the uniform operator topology, jointly in all the variables $t, \tau, s \in [0, T]$.

Proof: For $0 \leqslant t + \Delta t \leqslant T$, $\varepsilon \leqslant \tau + \Delta\tau \leqslant T$, and $0 \leqslant s + \Delta s \leqslant T$ we have the identity

$$A(t + \Delta t)\exp[-(\tau + \Delta\tau)A(s + \Delta s)] - A(t)\exp[-\tau A(s)]$$

$$= [A(t + \Delta t) - A(t)]A^{-1}(s + \Delta s) \cdot A(s + \Delta s)\exp[-(\tau + \Delta\tau)A(s + \Delta s)]$$

$$+ A(t)(\exp[-\Delta\tau A(s + \Delta s)] - \exp[-0 \cdot A(s + \Delta s)])$$

$$\times A^{-2}(s + \Delta s) \cdot A^2(s + \Delta s)\exp[-\tau A(s + \Delta s)]$$

$$+ A(t)(\exp[-\tau A(s + \Delta s)] - \exp[-\tau A(s)])$$

$$\equiv I_1 + I_2 + I_3. \tag{3.3.26}$$

Hypothesis 3, together with (3.1.14), gives

$$\|I_1\| \leqslant C|\Delta t|^{\alpha} C/(\tau + \Delta\tau)\exp[-\delta(\tau + \Delta\tau)]$$

$$\leqslant C|\Delta t|^{\alpha}. \tag{3.3.27}$$

The relations (3.3.9) and (3.1.17), with $n = 2$, yield

$$\|I_2\| \leqslant C|\Delta\tau|\exp[-\delta\min(0, \Delta\tau)](C/\tau^2)\exp(-\delta\tau)$$

$$\leqslant C|\Delta\tau|. \tag{3.3.28}$$

Finally, from (3.3.6) we obtain

$$\|I_3\| \leqslant (C/\tau)|\Delta s|^{\alpha}\exp(-\delta\tau)$$

$$\leqslant C|\Delta s|^{\alpha}. \tag{3.3.29}$$

In view of (3.3.26)–(3.3.29) the result follows.

LEMMA 3.3.4. The operator-valued function $A(t)\exp[-\tau A(s)]A^{-1}(\xi)$ is strongly continuous in X in all the variables $t, \tau, s, \xi \in [0, T]$.

Proof: Because of the identity

$$A(t)\exp[-\tau A(s)]A^{-1}(\xi)$$
$$= A(t)A^{-1}(s)\exp[-\tau A(s)]A(s)A^{-1}(\xi), \qquad (3.3.30)$$

it suffices to show that the operator-valued functions $A(t)A^{-1}(s)$ and $A(s)A^{-1}(\xi)$ are uniformly continuous in X and that the function $\exp[-\tau A(s)]$ is strongly continuous in X. Notice that

$$A(t+\Delta t)A^{-1}(s+\Delta s) - A(t)A^{-1}(s)$$
$$= [A(t+\Delta t) - A(t)]A^{-1}(s+\Delta s)$$
$$+ A(t)A^{-1}(s)[A(s) - A(s+\Delta s)]A^{-1}(s+\Delta s)$$
$$= J_1 + J_2.$$

From Hypothesis 3 $\|J_1\| \leqslant C|\Delta t|^{\alpha}$. Also (3.1.15) and Hypothesis 3 show that $\|J_2\| \leqslant C|\Delta s|^{\alpha}$ and the uniform continuity of $A(t)A^{-1}(s)$ follows.

Next we prove that the function $\exp[-\tau A(s)]$ is strongly continuous in X. In fact

$$(\exp[-(\tau+\Delta\tau)A(s+\Delta s)] - \exp[-\tau A(s)])A^{-1}(0)$$
$$= (\exp[-(\tau+\Delta\tau)A(s+\Delta s)] - \exp[-\tau A(s+\Delta s)])A^{-1}(0)$$
$$+ (\exp[-\tau A(s+\Delta s)] - \exp[-\tau A(s)])A^{-1}(0)$$
$$= I_1 + I_2.$$

From (3.3.8) $\|I_1\| \leqslant C|\Delta\tau|$ and in view of (3.3.5) and the boundedness of $A^{-1}(0)$ we have $\|I_2\| \leqslant C|\Delta s|^{\alpha}$. Hence, for each $v \in D$ the function $\exp[-A(s)]v$ is continuous. Since by (3.1.13), $\exp[-\tau A(s)]$ is uniformly bounded and $\bar{D} = X$, it follows that $\exp[-\tau A(s)]$ is strongly continuous in X. The proof is complete.

PROBLEM 3.3.2.   For $0 < \varepsilon \leqslant \tau + \varepsilon \leqslant t \leqslant T$, the operator-valued functions

$$[A(\tau)-A(t)]\exp[-(t-\tau)A(\tau)], \qquad [A(\tau)-A(t)]\exp[-(t-\tau)A(t)],$$
$$\exp[-(t-\tau)A(\tau)], \qquad \exp[-(t-\tau)A(t)]$$

are uniformly continuous in the uniform operator topology (that is, in the norm of $B(X)$), jointly, in the variables t and τ.

[Hint: Use Lemma 3.3.3.]

PROBLEM 3.3.3. For $0 \leqslant \tau \leqslant t \leqslant T$, the operator-valued functions

$$\exp[-(t-\tau)A(\tau)], \qquad \exp[-(t-\tau)A(t)],$$
$$[A(\tau)-A(t)]\exp[-(t-\tau)A(\tau)]$$

are continuous in X, jointly, in the variables t and τ.

[Hint: Write $\exp[-(t-\tau)A(\tau)] = A(\tau)\exp[-(t-\tau)A(\tau)]A^{-1}(\tau)$ and use Lemma 3.3.4.]

LEMMA 3.3.5. For $0 \leqslant \tau < t \leqslant t+\Delta t \leqslant T$ and any $\eta \in [0, \alpha]$ the function $\phi_1(t, \tau) \equiv [A(\tau)-A(t)]\exp[-(t-\tau)A(\tau)]$ satisfies the inequality

$$\|\phi_1(t+\Delta t, \tau) - \phi_1(t, \tau)\| \leqslant C(t-\tau)^{\eta-1}(\Delta t)^{\alpha-\eta}\exp[-\delta(t-\tau)].$$

$$(3.3.31)$$

Proof: Let us set

$$\phi_1(t, \tau) = [A(\tau)-A(t)]A^{-1}(\tau)A(\tau)\exp[-(t-\tau)A(\tau)].$$

Then as a result of Hypothesis 3 and (3.1.14), it follows that

$$\|\phi_1(t, \tau)\| \leqslant C|t-\tau|^{\alpha-1}\exp[-\delta(t-\tau)]. \qquad (3.3.32)$$

Notice that

$$J \equiv \phi_1(t+\Delta t, \tau) - \phi_1(t, \tau)$$
$$= [A(t) - A(t+\Delta t)]\exp[-(t+\Delta t-\tau)A(\tau)]$$
$$\quad + [A(\tau)-A(t)](\exp[-(t+\Delta t-\tau)A(\tau)] - \exp[-(t-\tau)A(\tau)])$$
$$= J_1 + J_2.$$

By Hypothesis 3 and (3.1.16)

$$\|J_1\| = \|[A(t) - A(t+\Delta t)]A^{-1}(t)A(t)\exp[-(t+\Delta t-\tau)A(\tau)]\|$$
$$\leqslant C(\Delta t)^{\alpha}[C/(t+\Delta t-\tau)]\exp[-\delta(t+\Delta t-\tau)]$$
$$\leqslant C(\Delta t)^{\alpha}(t-\tau)^{-1}\exp[-\delta(t-\tau)] \qquad (3.3.33)$$

and

$$\|J_2\| = \|[A(\tau)-A(t)]A^{-1}(\tau)$$
$$\quad \times (A(\tau)\exp[-(t+\Delta t-\tau)A(\tau)] - A(\tau)\exp[-(t-\tau)A(\tau)])\|$$
$$\leqslant C(t-\tau)^{\alpha}([C/(t+\Delta t-\tau)]\exp[-\delta(t+\Delta t-\tau)]$$
$$\quad + C/(t-\tau)\exp[-\delta(t-\tau)])$$
$$\leqslant C(t-\tau)^{\alpha-1}\exp[-\delta(t-\tau)]. \qquad (3.3.34)$$

On the other hand, for $x \in X$

$$J_2 x = -[A(\tau) - A(t)](I - \exp[-\Delta t A(\tau)]) \exp[-(t-\tau)A(\tau)] x$$

$$= -[A(\tau) - A(t)] A^{-1}(\tau) \int_0^{\Delta t} \exp[-\xi A(\tau)] A^2(\tau)$$

$$\times \exp[-(t-\tau)A(\tau)] x \, d\xi$$

and consequently

$$\|J_2\| \leqslant C(t-\tau)^\alpha \int_0^{\Delta t} C \exp(-\delta\xi) [C/(t-\tau)^2] \exp[-\delta(t-\tau)] \, d\xi$$

$$\leqslant C(t-\tau)^{\alpha-2}(\Delta t) \exp[-\delta(t-\tau)]. \tag{3.3.35}$$

The relations (3.3.34) and (3.3.35) show that

$$\|J_2\| = \|J_2\|^{1-\alpha} \|J_2\|^\alpha$$

$$\leqslant C(t-\tau)^{(\alpha-1)(1-\alpha)} \exp[-\delta(1-\alpha)(t-\tau)](t-\tau)^{(\alpha-2)\alpha}$$

$$\times (\Delta t)^\alpha \exp[-\delta\alpha(t-\tau)]$$

$$= C(t-\tau)^{-1}(\Delta t)^\alpha \exp[-\delta(t-\tau)]. \tag{3.3.36}$$

Thus from (3.3.33) and (3.3.36) we derive

$$\|J\| \leqslant C(t-\tau)^{-1}(\Delta t)^\alpha \exp[-\delta(t-\tau)]. \tag{3.3.37}$$

Moreover, from the definition of J and (3.3.32) there results the inequality

$$\|J\| \leqslant C(t-\tau)^{\alpha-1} \exp[-\delta(t-\tau)]. \tag{3.3.38}$$

In view of (3.3.38) and (3.3.37) we finally obtain

$$\|J\| = \|J\|^{\eta/\alpha} \|J\|^{(\alpha-\eta)/\alpha}$$

$$\leqslant C(t-\tau)^{\eta(\alpha-1)/\alpha} \exp[-(\delta\eta/\alpha)(t-\tau)] C(t-\tau)^{(\eta-\alpha)/\alpha}(\Delta t)^{\alpha-\eta}$$

$$\times \exp([-\delta(\alpha-\eta)/\alpha](t-\tau))$$

$$= C(t-\tau)^{\eta-1}(\Delta t)^{\alpha-\mu} \exp[-\delta(t-\tau)].$$

The proof is complete.

LEMMA 3.3.6.   The following estimates hold:

(a) $$\left\| A(t) \int_\tau^t \exp[-(t-s)A(s)] \, ds \right\| \leqslant C. \tag{3.3.39}$$

(b) $\qquad \left\| A(t) \int_\tau^{t-\rho} \exp[-(t-s)A(s)]\, ds \right\| \leqslant C, \qquad \rho > 0 \qquad$ (3.3.40)

Proof: (a) Set

$$F(t,\tau)x = A(t) \int_\tau^t \exp[-(t-s)A(s)]\,x\,ds, \qquad x \in D.$$

Then from (3.3.1) and Fubini's theorem, one gets

$$
\begin{aligned}
F(t,\tau)x = {}& A(t) \int_\tau^t \exp[-(t-s)A(s)]\,x\,ds \\
& + A(t) \int_\tau^t \left[\int_s^t \exp[-(t-\xi)A(t)] \right. \\
& \qquad\qquad \left. \times [A(t)-A(s)]\exp[-(\xi-s)A(s)]\,x\,d\xi \right] ds \\
= {}& (I - \exp[-(t-\tau)A(t)]) \left[I - \int_\tau^t \phi_1(t,s)\,x\,ds \right] \\
& + A(t) \int_\tau^t \exp[-(t-\xi)A(t)] \\
& \qquad \times \left(\int_\xi^t \phi_1(t,s)\,x\,ds + \int_\tau^t [\phi_1(t-s) - \phi_1(\xi,s)]\,x \right) d\xi \\
& + A(t) \int_\tau^t \exp[-(t-\xi)A(t)][A(t)-A(\xi)]A^{-1}(\xi)\,F(\xi,\tau)x\,d\xi.
\end{aligned}
$$

We now use Hypothesis 3, (3.1.13), (3.1.14), (3.3.32), and (3.3.31) to obtain, after some manipulations, the estimate

$$\|F(t,\tau)x\| \leqslant C\|x\| + C \int_\tau^t |\tau - \xi|^{\alpha-1} \|F(\xi,\tau)x\|\,d\xi.$$

We then have by Gronwall's inequality

$$\|F(t,\tau)x\| \leqslant C\|x\|, \qquad x \in D.$$

Since D is dense, the estimate (3.3.39) follows.

(b)   As the constant C does not depend on τ and t, we get

$$\left\| A(t) \int_\tau^{t-\rho} \exp[-(t-s)A(s)]\, ds \right\|$$

$$= \left\| A(t) \int_\tau^t \exp[-(t-s)A(s)]\, ds - A(t) \int_{t-\rho}^t \exp[-(t-s)A(s)]\, ds \right\|$$

$$\leqslant C.$$

The proof is complete.

3.4. Existence of a Fundamental Solution

In this section we prove the existence of a fundamental solution for the evolution equation (3.1.2) under Hypotheses 1–3 which we shall assume without further mention. As we pointed out in the heuristic remarks in Section 3.2, we will first show that the Volterra integral equation (3.2.10) has a solution $\Phi(t, \tau)$ and then prove that with this $\Phi(t, \tau)$, formula (3.2.11) gives a fundamental solution of (3.1.2).

LEMMA 3.4.1.   The Volterra integral equation (3.2.10) has a solution $\Phi(t, \tau)$ with $0 \leqslant \tau < t \leqslant T$ given by (3.2.9) which is uniformly continuous, in the topology of $B(X)$, in t, τ for $0 \leqslant \tau < t-\varepsilon$, $0 < \varepsilon \leqslant t \leqslant T$ and satisfies the estimate

$$\|\Phi(t, \tau)\| \leqslant C|t-\tau|^{\alpha-1}. \tag{3.4.1}$$

Under the restriction (3.4.1) the solution $\Phi(t, \tau)$ is unique and satisfies the equation

$$\Phi(t, \tau) = \phi_1(t, \tau) + \int_\tau^t \phi_1(t, s)\Phi(s, \tau)\, ds. \tag{3.4.2}$$

Proof: From the results of Section 3.3, the kernel $\phi_1(t, \tau)$ of (3.2.10) is uniformly continuous in (t, τ) in the uniform topology, provided that $t-\tau \geqslant \varepsilon > 0$, and satisfies the estimate (3.3.32). It follows by induction that the function $\phi_k(t, \tau)$ defined by (3.2.8) is uniformly continuous in (t, τ) in the uniform operator topology for $t-\tau \geqslant \varepsilon$ and satisfies the estimate

$$\|\phi_k(t, \tau)\| \leqslant C^k|t-\tau|^{k\alpha-1}/\Gamma(k\alpha) \tag{3.4.3}$$

where $\Gamma(n)$ is the gamma function. Thus the integral (3.2.8) makes sense, and the series (3.2.9) converges uniformly for $t-\tau \geqslant \varepsilon$ to a uniformly continuous function $\Phi(t, \tau)$ such that

$$\|\Phi(t, \tau)\| \leqslant \sum_{k=1}^{\infty} C^k |t-\tau|^{k\alpha-1}/\Gamma(k\alpha)$$

$$= C|t-\tau|^{\alpha-1} \sum_{k=1}^{\infty} C^{k-1} |t-\tau|^{\alpha(k-1)}/\Gamma(k\alpha)$$

$$\leqslant C|t-\tau|^{\alpha-1} \sum_{k=1}^{\infty} (CT^\alpha)^{k-1}/\Gamma(k\alpha)$$

$$\leqslant C|t-\tau|^{\alpha-1},$$

in view of the fact that the last numerical series converges. On the strength of (3.4.3), it follows that

$$\int_\tau^t \Phi(t, s) \phi_1(s, \tau) \, ds = \sum_{k=1}^{\infty} \int_\tau^t \phi_k(t, s) \phi_1(s, \tau) \, ds$$

$$= \sum_{k=1}^{\infty} \phi_{k+1}(t, \tau)$$

$$= \Phi(t, \tau) - \phi_1(t, \tau).$$

Hence, $\Phi(t, \tau)$ satisfies (3.2.10) and the estimate (3.4.1). Let $\Phi_1(t, \tau)$ and $\Phi_2(t, \tau)$ be two solutions of (3.2.10) satisfying (3.4.1). Then

$$\|\Phi_1(t, \tau) - \Phi_2(t, \tau)\| \leqslant \int_\tau^t \|\Phi_1(t, s) - \Phi_2(t, s)\| \, \|\phi_1(t, s)\| \, ds$$

which implies by Gronwall's inequality

$$\Phi_1(t, \tau) \equiv \Phi_2(t, \tau).$$

Finally we establish (3.4.2). Multiplying (3.2.10) on the left by $\phi_1(s, t)$ and integrating from τ to s, we get, using Fubini's theorem,

$$\int_\tau^s \phi_1(s, t) \Phi(t, \tau) \, dt$$

$$= \int_\tau^s \phi_1(s, t) \phi_1(t, \tau) \, dt + \int_\tau^s \phi_1(s, t) \left[\int_\tau^t \Phi(t, \xi) \phi_1(\xi, \tau) \, d\xi \right] dt$$

$$= \phi_2(s, \tau) + \int_\tau^s \left[\int_\xi^s \phi_1(s, t) \Phi(t, \xi) \, dt \right] \phi_1(\xi, \tau) \, d\xi.$$

Thus

$$\Psi(t, \tau) \equiv \int_\tau^t \phi_1(t, s) \Phi(s, \tau) \, ds$$

satisfies the equation

$$\Psi(t,\tau) = \phi_2(t,\tau) + \int_\tau^t \Psi(t,s)\phi_1(s,\tau)\,ds$$

and consequently, using the uniqueness of solution of (3.2.10),

$$\Psi(t,\tau) = \sum_{k=2}^\infty \phi_k(t,\tau) = \Phi(t,\tau) - \phi_1(t,\tau).$$

Equation (3.4.2) is therefore established. The proof is complete.

Equation (3.4.2) and Lemma 3.3.5 can be used to prove the following smoothness property of $\Phi(t,\tau)$.

LEMMA 3.4.2.   For $0 \leqslant \tau < t \leqslant t+\Delta t \leqslant T$ and any $\eta \in (0,\alpha]$

$$\|\Phi(t+\Delta t,\tau) - \Phi(t,\tau)\| \leqslant C(\Delta t)^{\alpha-\eta}(t-\tau)^{\eta-1} \tag{3.4.4}$$

where the constant C depends on η.

Proof: From (3.4.2) it follows that

$$\begin{aligned}
\Phi(t+\Delta t,\tau) - \Phi(t,\tau) &= \phi_1(t+\Delta t,\tau) - \phi_1(t,\tau) \\
&\quad + \int_t^{t+\Delta t} \phi_1(t+\Delta t,s)\Phi(s,\tau)\,ds \\
&\quad + \int_\tau^t [\phi_1(t+\Delta t,s) - \phi_1(t,s)]\Phi(s,\tau)\,ds.
\end{aligned} \tag{3.4.5}$$

In view of (3.3.31), (3.3.32), and (3.4.1) we obtain

$$\begin{aligned}
\|\Phi(t&+\Delta t,\tau) - \Phi(t,\tau)\| \\
&\leqslant C(t-\tau)^{\eta-1}(\Delta t)^{\alpha-\eta}\exp[-\delta(t-\tau)] \\
&\quad + \int_t^{t+\Delta t} C\exp[-\delta(t+\Delta t-s)](t+\Delta t-s)^{\alpha-1}(s-\tau)^{\alpha-1}\,ds \\
&\quad + \int_\tau^t C(t-s)^{\eta-1}(\Delta t)^{\alpha-\eta}\exp[-\delta(t-s)](s-\tau)^{\alpha-1}\,ds \\
&\leqslant C(\eta)(\Delta t)^{\alpha-\eta}(t-\tau)^{\eta-1}.
\end{aligned}$$

The proof is complete.

PROBLEM 3.4.1. For $0 \leqslant \tau < t \leqslant t + \Delta t \leqslant T$ and any $\eta \in (0, \alpha]$

$$\|[\Phi(t+\Delta t, \tau) - \Phi(t, \tau)] A^{-1}(\tau)\| \leqslant C(\Delta t)^{\alpha-\eta}(t-\tau)^{\eta} \qquad (3.4.6)$$

where the constant C depends on η.

[Hint: Use (3.4.2).]

The following lemma is the existence part of Theorem 3.2.1.

LEMMA 3.4.3. The operator $U(t, \tau)$ defined by (3.2.11) is a fundamental solution of (3.1.2).

Proof: From the results of Section 3.3 and Lemma 3.4.1 it follows that the operator function $U(t, \tau)$ is uniformly continuous in the variables t, τ for $t > \tau$, and is strongly continuous when $t \geqslant \tau$. It is also obvious that $U(\tau, \tau) = I$. Next, we shall prove that $U(t, \tau)$ is strongly continuously differentiable in t for $t > \tau$, that the range of $U(t, s)$ is in D, and that (3.1.5) holds. Let $t - \tau \geqslant \varepsilon > 0$, $0 < \rho < \varepsilon$, and $x \in X$. Define

$$U_\rho(t, \tau) x = \exp[-(t-\tau)A(\tau)] x + \int_\tau^{t-\rho} \exp[-(t-s)A(s)] \Phi(s, \tau) x \, ds. \qquad (3.4.7)$$

$U_\rho(t, \tau)$ is continuously differentiable in t and using (3.4.2) we get

$$\begin{aligned}
\partial U_\rho(t, \tau) x / \partial t = {}& -A(t)\exp[-(t-\tau)A(\tau)] x - A(t) \\
& \times \int_\tau^{t-\rho} \exp[-(t-s)A(s)] \Phi(s, \tau) x \, ds \\
& + \int_{t-\rho}^t \phi_1(t, s) \Phi(s, \tau) x \, ds \\
& + \exp[-\rho A(t-\rho)] [\Phi(t-\rho, \tau) - \Phi(t, \tau)] x \\
& + (\exp[-\rho A(t-\rho)] - I) \Phi(t, \tau) x \\
= {}& J_1 + J_2 + J_3 + J_4 + J_5 .
\end{aligned} \qquad (3.4.8)$$

In view of (3.3.32), (3.4.1), and (3.4.4)

$$\|J_3\| \leqslant C\rho^\alpha/\varepsilon^{1-\alpha} \|x\| \qquad \text{and} \qquad \|J_4\| \leqslant C\rho^{\alpha-\eta}/\varepsilon^{1-\eta} \|x\|,$$
$$0 < \eta < \alpha, \quad t - \tau \geqslant \varepsilon$$

so that J_3 and J_4 converge uniformly to zero as $\rho \to 0$. Also $J_5 \to 0$ as $\rho \to 0$, uniformly with respect to $t \geqslant \tau + s$, because $\exp[-\rho A(t-\rho)] x \to x$ as $\rho \to 0$ and the function $\Phi(t,\tau) x$ is uniformly continuous in (t,τ) for $t \geqslant \tau + \varepsilon$. We now claim that J_2 converges as $\rho \to 0$, uniformly with respect to $t \geqslant \tau + \varepsilon$. Assuming this claim it then follows from Theorem 1.3.5 that $\int_\tau^t \exp[-(t-s)A(s)] \Phi(s,\tau) x \, ds \in D$ and as $\rho \to 0$

$$J_2 \to -A(t) \int_\tau^t \exp[-(t-s)A(s)] \Phi(s,\tau) x \, ds.$$

Hence $U(t,\tau) x \in D$ and as $\rho \to 0$, we obtain from (3.4.8)

$$\partial U_\rho(t,\tau)/\partial t \to -A(t) U(t,\tau) x \qquad (3.4.9)$$

uniformly in (t,τ) for $t \geqslant \tau + \varepsilon$. On the other hand, $U_\rho(t,\tau) x \to U(t,\tau) x$ uniformly in (t,τ). Therefore by a standard argument, $[\partial U(t,\tau)/\partial t] x$ exists, is continuous in (t,τ) for $t > \tau$, and

$$[\partial U(t,\tau)/\partial t] x = \lim_{\rho \to 0} \partial U_\rho(t,\tau)/\partial t \, x$$
$$= -A(t) U(t,\tau) x.$$

It remains to establish the claim. We write

$$A(t) \int_\tau^{t-\rho} \exp[-(t-s)A(s)] \Phi(s,\tau) x \, ds$$

$$= A(t) \int_\tau^{t-\rho} \exp[-(t-s)A(s)] [\Phi(s,\tau) - \Phi(t,\tau)] x \, ds$$

$$+ A(t) \int_\tau^{t-\rho} \exp[-(t-s)A(s)] \Phi(t,\tau) x \, ds$$

$$= I_1 + I_2.$$

Because of (3.4.6) it suffices to prove the uniform convergence of I_2. Since $\Phi(t,\tau) x$ is continuous in t for $t \geqslant \tau + \varepsilon$ it is enough to establish the uniform convergence of the operator

$$A(t) \int_\tau^{t-\rho} \exp[-(t-s)A(s)] x \, ds, \qquad x \in X. \qquad (3.4.10)$$

This is obvious for any $x \in D$. Since D is dense in X and by Lemma 3.3.6, the operator (3.4.10) is bounded and the final assertion follows. The proof is complete.

3.5. Uniqueness of the Fundamental Solution

In Lemma 3.4.3 we exhibited a fundamental solution $U(t, \tau)$ of (3.1.2), namely, the operator defined by (3.2.11). In this section we shall complete the proof of Theorem 3.2.1 by proving that (3.1.2) has a unique fundamental solution. To this end it suffices to prove the following lemma:

LEMMA 3.5.1. Let Hypotheses 1–3 be satisfied. Then for any $u_0 \in X$ and any $\tau \in [0, T)$ the abstract Cauchy problem

$$du/dt + A(t)u = 0, \qquad \tau < t \leqslant T \quad \text{and} \quad u(\tau) = u_0 \qquad (3.5.1)$$

has the unique solution $u(t) = U(t, \tau) u_0$ where $U(t, \tau)$ is any fundamental solution of (3.1.2).

Assume that Lemma 3.5.1 has been established. Then if $U_1(t, \tau)$ and $U_2(t, \tau)$ are two fundamental solutions of (3.1.2), we should have that for any $u_0 \in X$ the functions $U_1(t, \tau) u_0$ and $U_2(t, \tau) u_0$ are both solutions of (3.5.1). Because of uniqueness it follows that $U_1(t, \tau) u_0 = U_2(t, \tau) u_0$, and consequently $U_1(t, \tau) = U_2(t, \tau)$.

The following two lemmas are needed in the proof of Lemma 3.5.1.

LEMMA 3.5.2. For $0 \leqslant t \leqslant T$ the operators $A(t)$ are bounded and satisfy hypotheses 1–3. Let $f \in C[[0, T], X]$. Then the Cauchy problem

$$du/dt + A(t)u = f(t), \qquad \tau < t \leqslant T \quad \text{and} \quad u(\tau) = 0 \qquad (3.5.2)$$

has a unique solution.

Proof: Let $W(t, \tau)$ be a fundamental solution of (3.1.2) corresponding to the bounded operators $A(t)$. Then

$$u(t) = \int_{\tau}^{t} W(t, s) f(s) \, ds$$

is a solution of (3.5.2) as one can verify by direct differentiation. To prove uniqueness, suppose that $f(t) = 0$ in (3.5.2). Then any solution satisfies the integral equation

$$u(t) = - \int_{\tau}^{t} A(s) u(s) \, ds.$$

It follows that

$$\|u(t)\| \leqslant \int_\tau^t \|A(s)\| \, \|u(s)\| \, ds$$

$$\leqslant C \int_\tau^t \|u(s)\| \, ds, \qquad u(\tau) = 0$$

and by Gronwall's inequality $u(t) \equiv 0$. The proof is complete.

LEMMA 3.5.3. For any $x \in X$, the function $A(t)\,U(t,\tau)\,A^{-1}(\tau)x$ is uniformly continuous in (t,τ) for $0 \leqslant \tau < t \leqslant T$. Moreover, for all (t,τ) with $0 \leqslant \tau < t \leqslant T$

$$\|A(t)\,U(t,\tau)\,A^{-1}(\tau)\| \leqslant C. \tag{3.5.3}$$

Proof: Define the functions

$$W(t,\tau) = A(t)\,U(t,\tau)\,A^{-1}(\tau), \qquad 0 \leqslant \tau < t \leqslant T$$

and

$$F(s) = \exp[-(t-s)A(t)]\,U(s,\tau)\,A^{-1}(\tau), \qquad \tau \leqslant s \leqslant t.$$

Then, $F(s)$ is continuously differentiable and

$$\begin{aligned}
F'(s) &= \exp[-(t-s)A(t)]\,A(t)\,U(s,\tau)\,A^{-1}(\tau)\,x \\
&\quad - \exp[-(t-s)A(t)]\,A(s)\,U(s,\tau)\,A^{-1}(\tau)\,x \\
&= \exp[-(t-s)A(t)]\,[A(t)-A(s)]\,U(s,\tau)\,A^{-1}(\tau)\,x.
\end{aligned}$$

Multiplying both sides by $A(t)$ and integrating the result with respect to s from τ to t, we obtain

$$\begin{aligned}
W(t,\tau)x &= A(t)\exp[-(t-\tau)A(t)]\,A^{-1}(\tau)\,x \\
&\quad + \int_\tau^t A(t)\exp[-(t-s)A(t)]\,[A(t)-A(s)]\,U(s,\tau)\,A^{-1}(\tau)\,x \, ds \\
&= A(t)\exp[-(t-\tau)A(t)]\,A^{-1}(\tau)\,x \\
&\quad + \int_\tau^t A(t)\exp[-(t-s)A(t)]\,[A(t)-A(s)]\,A^{-1}(s)\,W(s,\tau)\,x \, ds \\
&\equiv I_1 + I_2. \tag{3.5.4}
\end{aligned}$$

By Lemma 3.3.4, I_1 is uniformly continuous in t, τ for $t > \tau$. Since $W(t,\tau)\,x$ is also uniformly continuous in (t,τ) for $t-\tau \geqslant \varepsilon > 0$ it remains to show that

I_2 converges uniformly to a limit as $t - \tau \to 0$. Notice that

$$\|A(t)\exp[-(t-s)A(t)][A(t) - A(s)]A^{-1}(s)\| \leqslant C/(t-s)C(t-s)^{\alpha}$$
$$= C(t-s)^{\alpha-1}. \quad (3.5.5)$$

Also

$$\|I_1\| = \|\exp[-(t-s)A(t)]A(t)A^{-1}(\tau)x\|$$
$$\leqslant C\|x\|. \quad (3.5.6)$$

Hence, from (3.5.4), (3.5.5), and (3.5.6), we obtain

$$\|W(t,\tau)x\| \leqslant C\|x\| + C\int_{\tau}^{t}(t-s)^{\alpha-1}\|W(t,\tau)x\|\,ds,$$

and consequently

$$\|W(t,\tau)x\| \leqslant C\|x\|, \quad (3.5.7)$$

which proves (3.5.3). Now (3.5.5) and (3.5.7) show that

$$\|I_2\| \leqslant \int_{\tau}^{t} C(t-s)^{\alpha-1}C\|x\|\,ds$$
$$= C(t-\tau)^{\alpha}\|x\| \to 0 \quad \text{as} \quad t - \tau \to 0,$$

uniformly in (t,τ). The proof is complete.

Proof of Lemma 3.5.1: By Lemma 3.5.2, Lemma 3.5.1 is true if $A(t)$ is bounded for $0 \leqslant t \leqslant T$. Clearly, $U(t,\tau)u_0$ is a solution of (3.5.1). We shall prove the uniqueness result of Lemma 3.5.1 for the case of unbounded operators $A(t)$ by approximating (3.5.1) with the problems

$$dv/dt + A_n(t)v = 0 \quad \text{and} \quad v(\tau) = u_0 \quad (3.5.8)$$

where $A_n(t)$ is the bounded operator given by

$$A_n(t) = A(t)[I + n^{-1}A(t)]^{-1}, \quad n = 1, 2, \ldots.$$

We shall first prove that for each $n = 1, 2, \ldots$ the operators $A_n(t)$ for $0 \leqslant t \leqslant T$ satisfy Hypotheses 1–3. Moreover, the constants of Hypotheses 1–3 are independent of n and t, τ. Clearly, $A_n(t)$ is bounded for each n and Hypothesis 1 is satisfied. To prove Hypothesis 2 observe that for $\lambda \in S$

$$A_n(t) + \lambda I = (n+\lambda)/(n)[n\lambda/(n+\lambda)I + A(t)][I + n^{-1}A(t)]^{-1}.$$

Therefore

$$[A_n(t) + \lambda I]^{-1} = (n+\lambda)^{-1}I + [n^2/(n+\lambda)^2](A(t) + [(n\lambda/n+\lambda)]I)^{-1}$$

is bounded and

$$\|[A_n(t)+I]^{-1}\| \leqslant |n+\lambda|^{-1} + |n+\lambda|^{-1}[C/(|1+\lambda/n|+|\lambda|)]$$
$$\leqslant C/(1+|\lambda|).$$

Thus, Hypothesis 2 is valid. Finally we shall verify Hypothesis 3' which is equivalent to Hypothesis 3. We have

$$A_n(t)A_n^{-1}(s) = A_n(t)[A^{-1}(s)+n^{-1}]$$
$$= A(t)[I+n^{-1}A(t)]^{-1}A^{-1}(s) + n^{-1}A(t)[I+n^{-1}A(t)]^{-1}$$
$$= [I+n^{-1}A(t)]^{-1}A(t)A^{-1}(s) + I - [I+n^{-1}A(t)]^{-1}$$
$$= I + [I+n^{-1}A(t)]^{-1}[A(t)A^{-1}(s)-I].$$

Hence

$$\|[A_n(t)-A_n(s)]A_n^{-1}(s)\| \leqslant \|[I+n^{-1}A(t)]^{-1}\| \, \|[A(t)-A(s)]A^{-1}(s)\|$$
$$\leqslant C|t-s|^\alpha.$$

Next, let $U_n(t,\tau)$ be the fundamental solution corresponding to

$$dv/dt + A_n(t)v = 0, \qquad 0 \leqslant t \leqslant T,$$

as it was constructed in Section 3.4.

Then

$$\|U_n(t,\tau)\| \leqslant C + \int_\tau^t C\|\Phi_n(s,\tau)\| \, ds$$
$$\leqslant C + C\int_\tau^t (s-\tau)^{\alpha-1} \, ds$$
$$\leqslant C \qquad\qquad\qquad\qquad (3.5.9)$$

where C is independent of n.

Let $v_n(t)$ be the unique solution of (3.5.8) and $v(t)$ be any continuously differentiable solution of (3.5.1). The function $w_n(t) \equiv v(t)-v_n(t)$ satisfies the equation

$$dw_n/dt + A_n(t)w_n = [A_n(t)-A(t)]v(t) \qquad (3.5.10)$$

with

$$w_n(\tau) = 0. \qquad\qquad\qquad (3.5.11)$$

By Lemma 3.5.2 it follows that

$$w_n(t) = \int_\tau^t U_n(t,s)[A_n(s)-A(s)]v(s) \, ds \qquad (3.5.12)$$

is the unique solution of (3.5.10) and (3.5.11). We wish to show that $w_n(t) \to 0$ as $n \to \infty$, that is, $\lim_{n \to \infty} v_n(t) = v(t)$ which will imply that $v(t)$ is unique as being the limit of the unique solutions $v_n(t)$ of (3.5.8). As $n \to \infty$ the operators $[A_n(s) - A(s)] A^{-1}(s)$ converge to zero, uniformly with respect to s. Indeed, they are uniformly bounded

$$\|[A_n(s) - A(s)] A^{-1}(s)\| = \|[I + n^{-1} A(s)]^{-1} - I\|$$
$$\leqslant C + 1 = C$$

and for any $x \in D$

$$\|[A_n(s) - A(s)] A^{-1}(s) x\| = \|[I + n^{-1} A(s)]^{-1} x - x\|$$
$$= \|n^{-1} [I + n^{-1} A(s)]^{-1} A(s) A^{-1}(0) A(0) x\|$$
$$\leqslant n^{-1} C [C \|A(0) x\|] \to 0 \quad \text{as} \quad n \to \infty,$$

uniformly with respect to s.

Taking $x = A(s) v(s)$ which is continuous in s for $\tau \leqslant s \leqslant T$ it follows from (3.5.11) that $w_n(t) \to 0$ as $n \to \infty$. Finally, let $v(t)$ be a solution of (3.5.1) which is not necessarily continuously differentiable near $t = \tau$. For any $\tau < s < T$, $v(t)$ is continuously differentiable solution of (3.5.1) for $s \leqslant t \leqslant T$. From the identity

$$U(t, s) v(s) = U(t, s) A^{-1}(s) \cdot A(s) v(s)$$

and Lemma 3.5.3 it follows that the solution $U(t, s) v(s)$ is also continuously differentiable for $s \leqslant t \leqslant T$ with $U(s, s) v(s) = v(s)$. By the uniqueness result we have proved so far we conclude that $v(t) = U(t, s) v(s)$. Taking limits as $s \to \tau$ we get $v(t) = U(t, \tau) u_0$. The proof is complete.

COROLLARY 3.5.1.   For $0 \leqslant s \leqslant \tau \leqslant t \leqslant T$ the following identity holds:

$$U(t, \tau) U(\tau, s) = U(t, s). \tag{3.5.13}$$

Proof: For any $x \in X$, $U(t, s) x$ is the unique solution of (3.1.2) through (s, x). At time τ this solution goes through $(\tau, U(\tau, s) x)$. On the other hand, $U(t, \tau) U(\tau, s) x$ is the unique solution of (3.1.2) through $(\tau, U(\tau, s) x)$ and consequently coincides with $U(t, s) x$. The proof is complete.

PROBLEM 3.5.1.   Verify (3.2.4).

3.6. Solution of the Abstract Cauchy Problem

The present section is devoted to the proof of Theorem 3.2.2. In Lemma 3.5.1 we proved that the homogeneous Cauchy problem (3.5.1) has the unique solution $U(t, \tau) u_0$. Therefore (3.2.1) will have been established if we prove that the function

$$w(t) = \int_0^t U(t, s) f(s) \, ds \tag{3.6.1}$$

satisfies the Cauchy problem

$$dw/dt + A(t)w = f(t), \qquad 0 < t \le T; \tag{3.6.2}$$

$$w(0) = 0. \tag{3.6.3}$$

We shall first prove the following lemmas:

LEMMA 3.6.1.   For $0 \le \tau < t < T$ the following estimates hold:

(a) $$\|U(t, \tau) - \exp[-(t-\tau)A(t)]\| \le C|t-\tau|^{\alpha}; \tag{3.6.4}$$

(b) $$\|A(t)[U(t, \tau) - \exp[-(t-\tau)A(t)]]\| \le C|t-\tau|^{\alpha-1}; \tag{3.6.5}$$

(c) $$\|A(t) U(t, \tau)\| \le C|t-\tau|^{-1}. \tag{3.6.6}$$

Proof: (a)   We have, because of (3.2.11),

$$U(t, \tau) - \exp[-(t-\tau)A(t)] = (\exp[-(t-\tau)A(\tau) - \exp[-(t-\tau)A(t)])$$

$$+ \int_\tau^t \exp[-(t-s)A(s)] \, \Phi(s, \tau) \, ds$$

$$\equiv I_1 + I_2. \tag{3.6.7}$$

To estimate I_1, we use the relation (3.3.5) and get

$$\|I_1\| \le C|t-\tau|^{\alpha} \exp[-\delta(t-\tau)]$$

$$\le C|t-\tau|^{\alpha}. \tag{3.6.8}$$

The inequality (3.1.13) together with (3.4.1) yields

$$\|I_2\| \le \int_\tau^t C \exp[-\delta(t-s)] \, C|\tau-s|^{\alpha-1} \, ds$$

$$\le C|t-\tau|^{\alpha}. \tag{3.6.9}$$

The desired estimate (3.6.4) follows as a consequence of (3.6.7), (3.6.8), and (3.6.9).

(b)   We have from (3.6.7)

$$A(t)(U(t,\tau) - \exp[-(t-\tau)A(t)]) = A(t)I_1 + A(t)I_2.$$

Using (3.3.6), we get

$$\|A(t)I_1\| \leqslant [C/(t-\tau)]\,|t-\tau|^{\alpha}\exp[-\delta(t-\tau)]$$

$$\leqslant C|t-\tau|^{\alpha-1}. \tag{3.6.10}$$

Notice that for $x \in X$

$$A(t)I_2 = \int_\tau^t A(t)\exp[-(t-s)A(s)][\Phi(s,\tau) - \Phi(t,\tau)]\,ds$$

$$+ A(t)\left(\int_\tau^t \exp[-(t-s)A(s)]\,ds\right)\Phi(t,\tau). \tag{3.6.11}$$

In view of (3.1.16), (3.4.4), (3.3.39), and (3.4.1), we obtain for any $\eta \in (0,\alpha)$

$$\|A(t)I_2\| \leqslant \int_\tau^t [C/(t-s)]\,C(t-s)^{\alpha-\eta}(s-\tau)^{\eta-1}\,ds + C[C(t-\tau)^{\alpha-1}]$$

$$\leqslant C(t-\tau)^{\alpha-1}, \tag{3.6.12}$$

and the inequality (3.6.5) is established.

(c)   By (3.6.5) and (3.1.14), we get

$$\|A(t)U(t,\tau)\| \leqslant \|A(t)\exp[-(t-\tau)A(t)]\| + C|t-\tau|^{\alpha-1}$$

$$\leqslant [C/(t-\tau)] + C|t-\tau|^{\alpha-1}$$

$$\leqslant C|t-\tau|^{-1},$$

which proves (3.6.6).

LEMMA 3.6.2.   For any $x \in X$ with $0 \leqslant t \leqslant T$

$$\lim_{\Delta t \to 0+} ([U(t+\Delta t, t) - I]/\Delta t)\,A^{-1}(t)x = -x. \tag{3.6.13}$$

Proof: The function

$$F(s) = \exp[-(t-s)A(t)]\,U(s,\tau)y, \qquad y \in X, \quad \tau \leqslant s \leqslant t,$$

is continuously differentiable and

$$F'(s) = \exp[-(t-s)A(t)][A(t) - A(s)]\,U(s,\tau)y. \tag{3.6.14}$$

Integrating (3.6.14) from τ to t, we obtain

$$U(t, \tau) y = \exp[-(t-\tau) A(t)] y$$

$$+ \int_\tau^t \exp[-(t-s) A(t)] [A(t) - A(s)] U(s, \tau) y \, ds.$$

Setting $y = A^{-1}(t) x$, it follows that

$$([U(t+\Delta t, t) - I]/\Delta t) A^{-1}(t) x$$

$$= [(\exp[-\Delta t A(t+\Delta t)] - I)/\Delta t] A^{-1}(t) x$$

$$+ (\Delta t)^{-1} \int_t^{t+\Delta t} \exp[-(t+\Delta t - s) A(t+\Delta t)]$$

$$\times [A(t+\Delta t) - A(s)] A^{-1}(s) \cdot A(s) U(s, t) A^{-1}(t) x \, ds$$

$$\equiv J_1 + J_2.$$

By (3.1.13), Hypothesis 3, and (3.5.3), one gets

$$\|J_2\| \leqslant (\Delta t)^{-1} \int_t^{t+\Delta t} CC(t+\Delta t - s)^\alpha \, ds$$

$$\leqslant C(\Delta t)^\alpha \to 0 \qquad \text{as} \quad \Delta t \to 0_+. \tag{3.6.15}$$

On the other hand, since $y \in D$, we have

$$\exp[-\Delta t A(t+\Delta t)] y - y = - \int_0^{\Delta t} \exp[-\sigma A(t+\Delta t)] A(t+\Delta t) y \, d\sigma.$$

and as a result

$$J_1 = -(\Delta t)^{-1} \int_0^{\Delta t} \exp[-\sigma A(t+\Delta t)] A(t+\Delta t) A^{-1}(t) x \, d\sigma \to -x$$

$$\text{as} \quad \Delta t \to 0_+.$$

The proof is therefore complete.

PROBLEM 3.6.1.   Prove that for every $x \in X$, $\int_0^t U(t, s) x \, ds \in D$.

Now we are ready to show that (3.6.1) satisfies (3.6.2) and (3.6.3). Utilizing (3.5.13) we formally obtain

$[w(t+\Delta t) - w(t)]/\Delta t$

$$= (\Delta t)^{-1} \int_0^{t+\Delta t} U(t+\Delta t, s)f(s)\, ds - (\Delta t)^{-1} \int_0^t U(t, s)f(s)\, ds$$

$$= (\Delta t)^{-1} \int_t^{t+\Delta t} U(t+\Delta t, s)f(s)\, ds$$

$$+ (\Delta t)^{-1} \int_0^t [U(t+\Delta t, s) - U(t, s)]f(s)\, ds$$

$$= (\Delta t)^{-1} \int_t^{t+\Delta t} U(t+\Delta t, s)f(s)\, ds$$

$$+ ([U(t+\Delta t, t) - I]/\Delta t) \int_0^t U(t, s)f(s)\, ds$$

$$= (\Delta t)^{-1} \int_t^{t+\Delta t} U(t+\Delta t, s)f(s)\, ds$$

$$+ ([U(t+\Delta t, t) - I]/\Delta t) A^{-1}(t) \left[A(t) \int_0^t U(t, s)[f(s)-f(t)]\, ds \right.$$

$$+ A(t) \int_0^t (U(t, s) - \exp[-(t-s)A(t)])f(t)\, ds$$

$$+ \left. [I - \exp[-tA(t)]]f(t) \right]. \tag{3.6.16}$$

In obtaining (3.6.16) we have made use of the identity

$$\int_0^t \exp[-(t-s)A(t)]f(t)\, ds = -A^{-1}(t)[I - \exp[-tA(t)]]f(t).$$

In view of Lemma 3.6.1, Problem 3.6.1, and Hypothesis 4 all the terms in (3.6.16) make sense and consequently (3.6.16) is valid. Next, taking limits on both sides as $\Delta t \to 0_+$ and using Lemma 3.6.2 we conclude that $dw(t)/dt$ exists and

$$dw(t)/dt = f(t) - A(t) \int_0^t U(t, s)[f(s)-f(t)]\, ds$$

$$- A(t) \int_0^t (U(t, s) - \exp[-(t-s)A(t)])f(t)\, ds$$

$$- [I - \exp[(-tA(t)]]f(t). \tag{3.6.17}$$

Hence

$$dw(t)/dt = f(t) - A(t) \int_0^t U(t,s)f(s)\,ds$$

$$= f(t) - A(t)w(t).$$

Finally, one can verify that all the terms in (3.6.17) are continuous functions of t and consequently the solution $w(t)$ of (3.6.2) and (3.6.3) is continuously differentiable. The proof of Theorem 3.2.2 is therefore complete.

3.7. Differentiability of Solutions

In this section we shall prove Theorem 3.2.3 which asserts that under Hypotheses 1–3, 5, and 6 the solution $u(t)$ of the abstract Cauchy problem (3.1.3) and (3.1.4) is $(k+1)$-times (strongly) continuously differentiable on $[\varepsilon, T]$ for any $\varepsilon > 0$. We need the following lemmas.

LEMMA 3.7.1. Assume that Hypothesis 6 holds and that the operator $A(0)A^{-1}(t)$ is bounded for each $t \in [0, T]$. Then for any $x \in X$ the function $A(0)A^{-1}(t)x$ has continuous derivatives

$$(d^j/dt^j)[A(0)A^{-1}(t)x] \equiv A(0)[A^{-1}(t)]^{(j)}x, \qquad 1 \leqslant j \leqslant k.$$

Furthermore, the operators $A(0)[A^{-1}(t)]^{(j)}$ are uniformly bounded for $t \in [0, T]$.

Proof: Set $B(t) = A(t)A^{-1}(0)$ and $B_h(t) = [B(t+h) - B(t)]/h$. Then $B^{-1}(t) = A(0)A^{-1}(t)$ and $B(t+h)y - B(t)y = hB_h(t)y$ where $\|B_h(t)y\| \leqslant C$, C being independent of h. Multiplying both sides of this equation by $B^{-1}(t)$ on the left and taking $y = B^{-1}(t+h)x$ we obtain

$$B^{-1}(t+h)x - B^{-1}(t)x = -hB^{-1}(t)B_h(t)B^{-1}(t+h)x. \qquad (3.7.1)$$

From (3.7.1), we observe that

$$\|B^{-1}(t+h)x - B^{-1}(t)x\| \leqslant C|h| \|B^{-1}(t+h)\|,$$

which proves that $B^{-1}(t)x$ is continuous and

$$(d/dt)B^{-1}(t)x = -B^{-1}(t)B'(t)B^{-1}(t)x.$$

From this identity we see that the first k derivatives of $B^{-1}(t)x = A(0)A^{-1}(t)x$ exist, are continuous, and the operators $A(0)[A^{-1}(t)]^{(j)}$ are uniformly bounded for $1 \leqslant j \leqslant k$. The proof is complete.

LEMMA 3.7.2. Let $g(s)$ be a continuous function for $\tau \leqslant s \leqslant T$. Then the function $\int_{\tau}^{t} U(t, s) g(s) \, ds$ is uniformly Hölder continuous in t for $\tau \leqslant t \leqslant T$ with any exponent $\gamma \in (0, 1)$.

Proof: Let ϕ be a (real) bounded linear functional and $0 < \tilde{h} < h$. Employing the mean value theorem and (3.6.6) we notice that

$$|\phi [U(t+h, s) x] - \phi [U(t, s) x]| \leqslant h |\phi [(d/dt) U(t+\tilde{h}, s) x]|$$

$$\leqslant h \|\phi\| \|x\| \|(d/dt) U(t+\tilde{h}, s)\|$$

$$\leqslant C \|\phi\| \|x\| h |t-s|^{-1}.$$

This inequality implies that for $0 \leqslant h \leqslant |t-s|$

$$\|U(t+h, s) - U(t, s)\| \leqslant Ch^{\gamma}/|t-s|^{\gamma}. \tag{3.7.2}$$

Now observe that for $0 < h < t - \tau$ with $h < 1$

$$\int_{\tau}^{t+h} U(t+h, s) g(s) \, ds - \int_{\tau}^{t} U(t, s) g(s) \, ds$$

$$= \int_{t-h}^{t+h} U(t+h, s) g(s) \, ds - \int_{t-h}^{t} U(t, s) g(s) \, ds$$

$$+ \int_{\tau}^{t-h} [U(t+h, s) - U(t, s)] g(s) \, ds. \tag{3.7.3}$$

After some computations the relations (3.7.3) and (3.7.2) yield the desired conclusion.

Proof of Theorem 3.2.3: Let $u(t)$ be the solution of (3.1.3) and (3.1.4). Set $u_h(t) = [u(t+h) - u(t)]/h$. By (3.1.3) it follows that for any $0 < t < t+h \leqslant T$ $du_h(t)/dt + A(t) u_h(t) = ([f(t+h) - f(t)]/h) - ([A(t+h) - A(t)]/h) u(t+h)$. Making use of formula (3.2.1), we find that for any $\tau \in (0, t)$

$$u_h(t) = U(t, \tau) u_h(\tau) + \int_{\tau}^{t} U(t, s)$$

$$\times [([f(s+h) - f(s)]/h) - ([A(s+h) - A(s)]/h) u(s+h)] \, ds. \tag{3.7.4}$$

Clearly, $[f(s+h) - f(s)]/h \to f'(s)$ as $h \to 0$, uniformly with respect to $s \in [\tau, t]$. Moreover

$$([A(s+h) - A(s)]/h) u(s+h)$$

$$= ([A(s+h) A^{-1}(0) - A(s) A^{-1}(0)]/h) A(0) A^{-1}(s+h) A(s+h) u(s+h).$$

Using Hypothesis 6, Lemma 3.7.1, and the uniform continuity of $A(t)u(t)$ for $\varepsilon \leqslant t \leqslant T$ with $\varepsilon > 0$, we obtain

$$\lim_{h \to 0} ([A(s+h) - A(s)]/h) u(s+h) = A'(s)u(s)$$

uniformly with respect to $s \in [\tau, t]$. (Here, $A'(s)$ stands for $A'(s)A^{-1}(0) \cdot A(0)$.) Now taking limits as $h \to 0$ in (3.7.4) we find

$$u'(t) = U(t,\tau)u'(\tau) + \int_\tau^t U(t,s)[f'(s) - A'(s)u(s)] \, ds. \qquad (3.7.5)$$

In view of Lemma 3.7.2 the integral in (3.7.5) is uniformly Hölder continuous with any exponent $\gamma \in (0,1)$. Hence, the same is true for $u'(t)$ in $[\tau', T]$ for $\tau' > \tau$. Since τ is an arbitrary point in $(0, T]$, $u'(t)$ is uniformly Hölder continuous (with exponent γ) in $[\varepsilon, T]$ for any $\varepsilon > 0$. From (3.7.5) and the results of Section 3.6 it follows that $u'(t)$ is continuously differentiable in $[\tau', T]$ for any $\tau' > \tau$, and

$$d^2u(t)/dt^2 + A(t)\,du(t)/dt = f'(t) - A'(t)u(t). \qquad (3.7.6)$$

Since τ is an arbitrary point in $(0, T]$, $u''(t)$ is continuous in every interval $[\varepsilon, T]$ with $\varepsilon > 0$. Writing

$$A'(t)u(t) = A'(t)A^{-1}(0) \cdot A(0)A^{-1}(t) \cdot A(t)u(t)$$

and using Lemma 3.7.1 we see that $A'(t)u(t)$ is continuously differentiable. Applying the same arguments as before we find that for any $0 < \tau \leqslant t \leqslant T$

$$u''(t) = U(t,\tau)u''(\tau) + \int_\tau^t U(t,s)[f''(s) - A''(s)u(s) - 2A'(s)u'(s)] \, ds$$

where $A''(s)u(s)$ stands for

$$A''(s)A^{-1}(0) \cdot A(0)u(s) = [A''(s)A^{-1}(0)][A(0)A^{-1}(s)]A(s)u(s).$$

Again from Lemma 3.7.2 the function $u''(t)$ is uniformly Hölder continuous in every interval $[\varepsilon, T]$ for $\varepsilon > 0$. Hence $u'''(t)$ exists, is continuous in every interval $[\varepsilon, T]$ and

$$u'''(t) + A(t)u''(t) = -2A'(t)u'(t) - A''(t)u(t) + f''(t).$$

By induction, it now follows that

$$u^{(j+1)}(t) + A(t)u^{(j)}(t) = -\sum_{i=0}^{j-1} \binom{j}{i} A^{(j-i)}(t)u^{(i)}(t) + f^{(j)}(t).$$

The proof is complete.

3.8. Asymptotic Behavior

In this section we shall investigate the behavior as $t \to \infty$ of any solution $u(t)$ of (3.1.8). Our aim is to prove Theorem 3.2.4.

Proof of Theorem 3.2.4: Set

$$\eta(\mu) = \sup_{\substack{t > \tau \geq \mu \\ 0 \leq s < \infty}} \|[A(t) - A(\tau)] A^{-1}(s)\|. \tag{3.8.1}$$

By Hypothesis 9 $\eta(\mu) \to 0$ as $\mu \to \infty$. From (3.8.1) and Hypothesis 3 we get

$$\|[A(t) - A(\tau)] A^{-1}(s)\| \leq C[\eta(\mu)]^{1/2} |t - \tau|^{\alpha/2}, \qquad \mu \leq \tau < t, \quad s > 0. \tag{3.8.2}$$

Recalling the notation of Section 3.2 and using (3.1.14) and (3.8.1) we obtain for $t > \tau \geq \mu \geq 0$

$$\|\phi_1(t, \tau)\| \leq K[\eta(\mu)]^{1/2} |t - \tau|^{-1 + (\alpha/2)} \exp[-\delta(t - \tau)] \tag{3.8.3}$$

where K is a constant. Inductively, it follows that

$$\|\phi_k(t, \tau)\|$$
$$\leq (K[\eta(\mu)])^{1/2k} |t - \tau|^{(k\alpha/2) - 1} [\Gamma(\alpha/2)]^k [\Gamma(k\alpha/2)]^{-1} \exp[-\delta(t - \tau)]. \tag{3.8.4}$$

Hence, for any $\theta \in (0, \delta)$

$$\|\Phi(t, \tau)\| \leq C[\eta(\mu)]^{1/2} |t - \tau|^{-[1 - (\alpha/2)]} \exp[-\theta(t - \tau)], \qquad \mu \leq \tau < t, \tag{3.8.5}$$

where C is a generic constant independent of μ. In the light of (3.1.13), (3.1.14), (3.8.2), (3.8.3), and (3.8.5) one can establish the estimate

$$\|\Phi(t + \Delta t, \tau) - \Phi(t, \tau)\| \leq C[\eta(\mu)]^{1/2} (\Delta t)^{(\alpha/2) - \delta} |t - \tau|^{\delta - 1} \exp[-\theta(t - \tau)].$$

Then, analogously to (3.6.4), (3.6.5), and (3.6.6) one obtains, for $0 < \theta < \delta$ and $t > \tau > \mu \geq 0$, the estimates

$$\|U(t, \tau) - \exp[-(t - \tau) A(t)]\|$$
$$\leq C[\eta(\mu)]^{1/2} (t - \tau)^{\alpha/2} \exp[-\theta(t - \tau)] \tag{3.8.6}$$

$$\|A(t)(U(t, \tau) - \exp[-(t - \tau) A(t)])\|$$
$$\leq C[\eta(\mu)]^{1/2} (t - \tau)^{-1 + (\alpha/2)} \exp[-\theta(t - \tau)] \tag{3.8.7}$$

$$\|A(t) U(t, \tau)\| \leq C(t - \tau)^{-1} \exp[-\theta(t - \tau)]. \tag{3.8.8}$$

We shall now prove that

$$du/dt \to 0 \quad \text{as} \quad t \to \infty. \tag{3.8.9}$$

We have

$$u(t) = U(t,\tau)u(\tau) + \int_\tau^t U(t,s)f(s)\,ds \equiv U(t,\tau)u(\tau) + w(t) \tag{3.8.10}$$

where

$$
\begin{aligned}
dw/dt = f(t) - \Bigg(& \int_\tau^t A(t)\,U(t,s)\,[f(s)-f(t)]\,ds \\
& + \Bigg[\int_\tau^t A(t)(U(t,s) - \exp[-(t-s)A(t)])\,ds\Bigg]f(t) \\
& + (I - \exp[-(t-\tau)A(t)])f(t)\Bigg).
\end{aligned}
\tag{3.8.11}
$$

Set

$$\delta(\mu) = \sup_{\substack{t \geq \tau \geq \mu \\ 0 \leq s < \infty}} \|f(t)-f(\tau)\|.$$

Then by our assumptions, $\delta(\mu) \to 0$ as $\mu \to \infty$ and

$$\|f(s)-f(t)\| \leq C\delta(\mu)|t-s|^{\beta/2}, \quad s,t \geq \mu. \tag{3.8.12}$$

By (3.8.12) and (3.8.8), we obtain, for $\tau \geq \mu$

$$
\begin{aligned}
\Bigg\|\int_\tau^t A(t)\,U(t,s)\,&[f(s)-f(t)]\,ds\Bigg\| \\
&\leq C[\delta(\mu)]^{1/2}\int_\tau^t \exp[-\theta(t-s)]/[(t-s)^{-(1-\beta/2)}]\,ds \\
&= C[\delta(\mu)]^{1/2}.
\end{aligned}
\tag{3.8.13}
$$

Since $\sup_{t>0}\|f(t)\| < \infty$, we get from (3.8.7) the estimate

$$\Bigg\|\Bigg[\int_\tau^t A(t)(U(t,s) - \exp[-(t-s)A(t)])\,ds\Bigg]f(t)\Bigg\| \leq C[\eta(\mu)]^{1/2}. \tag{3.8.14}$$

In view of (3.8.13), (3.8.14), (3.1.13), and (3.8.11) we conclude that $\|dw/dt\|$ can be made arbitrarily small for t sufficiently large. The same is also true for $\|(d/dt)\,U(t,\tau)u(\tau)\|$. The claim (3.8.9) now follows from (3.8.10). By (3.8.9) and (3.1.8) we get

$$-A(t)u(t) + f(t) \to 0 \quad \text{as} \quad t \to \infty.$$

Since by hypothesis $f(t) \to f(\infty)$, it follows that

$$A(\infty)u(t) = A(\infty)A^{-1}(t) \cdot A(t)u(t) \to f(\infty)$$

and therefore

$$u(t) = A^{-1}(\infty)A(\infty)u(t) \to A^{-1}(\infty)f(\infty) \equiv u(\infty).$$

The proof is complete.

3.9. Notes

The results of this chapter are due to Sobolevski [65] and Tanabe [70, 71, 72, 73]. Here we follow very closely Sobolevski [65]. Section 3.8 is the work of Tanabe [73]. For further results and applications the reader is referred to Carroll [12], Friedman [23], and Kato [31]. For fractional powers see also Sobolevski [65] and Friedman [23]. The case of hyperbolic abstract Cauchy problems is treated in detail in the fundamental paper of Kato [30].

Chapter 4

Evolution Inequalities

4.0. Introduction

We present, in this chapter, a number of results concerning with lower bounds and uniqueness of solutions of evolution inequalities in a Hilbert space. Employing the method of scalar differential inequalities and using elementary methods, we first obtain lower bounds of solutions which are then profitably used to prove various kinds of uniqueness results.

For clarity, we focus our attention in Section 4.1 on a special evolution inequality with time independent evolution operator. After deriving quite general lower bounds we prove, as applications, interesting uniqueness results including a "unique continuation at infinity" theorem. We also deduce certain explicit lower bounds. In addition, we show that the solutions verify some convexity-like inequalities which in turn lead to the derivation of lower bounds. Section 4.2 deals with results of similar character. However, the evolution inequality considered here is rather general in many respects and also offers a much wider range of applicability to partial differential equations.

In Section 4.3 we give upper and lower bounds of solutions of a non-linear evolution inequality and prove a very general uniqueness result for such equations. In the entire discussion the operator A involved is assumed to be either symmetric or self-adjoint or its resolvent satisfies a growth condition. Finally, we study a parabolic partial differential inequality to illustrate specifically the meaning of the assumptions and the results obtained in this chapter.

4.1. Lower Bounds, Uniqueness, and Convexity (Special Results)

Let us consider the time independent evolution operator

$$Lu = u' - Au, \qquad ' = d/dt \tag{4.1.1}$$

in a Hilbert space H with inner product $(\cdot, \cdot)$ and norm $\|\cdot\|$. We shall assume that A on $D(A)$ is a linear symmetric operator (generally unbounded) in H, that is,

$$(Au, v) = (u, Av), \qquad u, v \in D(A).$$

Let $\phi: J \to R_+$ be a given measurable and locally bounded function defined on an interval J of the real line. Consider the evolution inequality

$$\|Lu(t)\| \leqslant \phi(t) \|u(t)\|, \qquad t \in J. \tag{4.1.2}$$

DEFINITION 4.1.1. A function $u \in C[J, H]$ is said to be *a solution* of the evolution inequality (4.1.2) if

(i) $\qquad\qquad u(t) \in D(A), \qquad t \in J;$

(ii)   the strong derivative $u'(t)$ exists and is piecewise continuous on J;
(iii) $u(t)$ satisfies the inequality (4.1.2) for all $t \in J$.

We pass now to our main results of this section.

THEOREM 4.1.1. Let A be a linear symmetric operator in H with domain $D(A)$. Let $u(t)$ be a solution of (4.1.2) such that $Au(t)$ is continuous for

$t \in J$. Then for $t, t_0 \in J$ the following lower bounds are valid:

$$\|u(t)\| \geqslant \|u(t_0)\| \exp\left(-\lambda_1 (t-t_0) - \int_{t_0}^{t} [\phi(s) + \tfrac{1}{2}(t-s)\phi^2(s)] \, ds \right),$$

$$t \geqslant t_0 \tag{4.1.3}$$

$$\|u(t)\| \geqslant \|u(t_0)\| \exp\left(-\lambda_2 (t_0-t) - \int_{t}^{t_0} [\phi(s) + \tfrac{1}{2}(s-t)\phi^2(s)] \, ds \right),$$

$$t \leqslant t_0 \tag{4.1.4}$$

where λ_1 and λ_2 are nonnegative constants depending on $u(t_0)$.

Proof: First assume that $u(t) \neq 0$ for $t \in J$. Set

$$m(t) = \|u(t)\|^2 \quad \text{and} \quad Q(t) = (Au(t), u(t))/\|u(t)\|^2.$$

In view of (4.1.1) and the symmetry of A we have

$$\begin{aligned} m'(t) &= (u'(t), u(t)) + (u(t), u'(t)) \\ &= 2\,\mathrm{Re}(u'(t), u(t)) \\ &= 2(Au(t), u(t)) + 2\,\mathrm{Re}(Lu(t), u(t)). \end{aligned} \tag{4.1.5}$$

Hence

$$|m'(t) - 2Q(t)m(t)| \leqslant 2|\mathrm{Re}(Lu(t), u(t))|. \tag{4.1.6}$$

Using Schwarz's inequality and (4.1.2) we get

$$|\mathrm{Re}(Lu(t), u(t))| \leqslant \phi(t)m(t).$$

This together with (4.1.6) yields

$$|m'(t) - 2Q(t)m(t)| \leqslant 2\phi(t)m(t),$$

which implies that

$$m'(t) \geqslant 2[Q(t) - \phi(t)]m(t), \qquad t \in J, \tag{4.1.7}$$

and

$$m'(t) \leqslant 2[Q(t) + \phi(t)]m(t), \qquad t \in J. \tag{4.1.8}$$

Since A is symmetric and $Au(t)$ is continuous in t

$$(d/dt)(Au(t), u(t)) = (Au'(t), u(t)) + (Au(t), u'(t)).$$

Indeed, from the definition of the derivative

$$(d/dt)(Au(t), u(t)) = \lim_{h \to 0} [(Au(t+h), u(t+h)) - (Au(t), u(t))]/h$$

$$= \lim_{h \to 0} [([Au(t+h) - Au(t)]/h, u(t+h))$$

$$+ (Au(t), [u(t+h) - u(t)]/h)]$$

$$= \lim_{h \to 0} [(u(t+h) - u(t)/h, Au(t+h))$$

$$+ (Au(t), u(t+h) - u(t)/h)]$$

$$= (u'(t), Au(t)) + (Au(t), u'(t))$$

$$= (Au'(t), u(t)) + (Au(t), u'(t)).$$

From this and (4.1.1) we see that

$$(d/dt)(Au(t), u(t)) = (Au'(t), u(t)) + (Au(t), u'(t))$$

$$= (u'(t), Au(t)) + (Au(t), u'(t))$$

$$= 2 \operatorname{Re}(Au(t), u'(t))$$

$$= 2 \operatorname{Re}(Au(t), Lu(t) + Au(t))$$

$$= 2 \|Au(t) + \tfrac{1}{2}Lu(t)\|^2 - \tfrac{1}{2}\|Lu(t)\|^2. \qquad (4.1.9)$$

It follows, using (4.1.5) and (4.1.9), that (writing u for $u(t)$)

$$Q'(t) = \|u\|^{-2}[2\|Au + \tfrac{1}{2}Lu\|^2 - \tfrac{1}{2}\|Lu\|^2]$$

$$- [(Au, u)/\|u\|^4][2(Au, u) + 2\operatorname{Re}(Lu, u)]$$

$$= \|u\|^{-4}(2\|Au + \tfrac{1}{2}Lu\|^2 \|u\|^2 - \tfrac{1}{2}\|Lu\|^2 \|u\|^2 - 2[\operatorname{Re}(Au + \tfrac{1}{2}Lu, u)]^2$$

$$+ \tfrac{1}{2}[\operatorname{Re}(Lu, u)]^2). \qquad (4.1.10)$$

Using Schwarz's inequality and (4.1.2) we get

$$Q'(t) \geq -\tfrac{1}{2}\phi^2(t), \qquad t \in J. \qquad (4.1.11)$$

This implies that for $t, t_0 \in J$

$$Q(t) \geq Q(t_0) - \tfrac{1}{2}\int_{t_0}^{t} \phi^2(s)\, ds, \qquad t \geq t_0, \qquad (4.1.12)$$

and

$$Q(t) \leq Q(t_0) + \tfrac{1}{2}\int_{t}^{t_0} \phi^2(s)\, ds, \qquad t \leq t_0. \qquad (4.1.13)$$

Now using (4.1.7) and (4.1.12) we obtain the differential inequality

$$m'(t) \geq 2\left[Q(t_0) - \phi(t) - \tfrac{1}{2}\int_{t_0}^{t}\phi^2(s)\,ds\right]m(t), \qquad t \geq t_0.$$

Similarly, from (4.1.8) and (4.1.13) we have

$$m'(t) \leq 2\left[Q(t_0) + \phi(t) + \tfrac{1}{2}\int_{t}^{t_0}\phi^2(s)\,ds\right]m(t), \qquad t \leq t_0.$$

Integrating these inequalities we obtain

$$m(t) \geq m(t_0)\exp 2\left[Q(t_0)(t-t_0) - \int_{t_0}^{t}[\phi(s) + \tfrac{1}{2}(t-s)\phi^2(s)]\,ds\right],$$
$$t \geq t_0$$

and

$$m(t_0) \leq m(t)\exp 2\left[Q(t_0)(t_0-t) + \int_{t}^{t_0}[\phi(s) + \tfrac{1}{2}(s-t)\phi^2(s)]\,ds\right],$$
$$t \leq t_0.$$

Since $m(t) = \|u(t)\|^2$, extracting square roots, we derive

$$\|u(t)\| \geq \|u(t_0)\|\exp\left[Q(t_0)(t-t_0) - \int_{t_0}^{t}[\phi(s) + \tfrac{1}{2}(t-s)\phi^2(s)]\,ds\right],$$
$$t \geq t_0. \tag{4.1.14}$$

and

$$\|u(t)\| \geq \|u(t_0)\|\exp\left[-Q(t_0)(t_0-t) - \int_{t}^{t_0}[\phi(s) + \tfrac{1}{2}(s-t)\phi^2(s)]\,ds\right],$$
$$t \leq t_0. \tag{4.1.15}$$

Setting $\lambda_1 = -\min(0, Q(t_0))$ and $\lambda_2 = \max(0, Q(t_0))$ the desired lower bounds follow. We have proved (4.1.3) and (4.1.4) assuming that $u(t) \neq 0$ for $t \in J$. If $u(t_0) = 0$ these bounds are clearly valid. If $u(t_0) \neq 0$, then $u(t)$ cannot vanish on J and the previous arguments are valid. Otherwise there exists an interval with one end point t_0, say $[t_0, t_1)$, such that $u(t) \neq 0$ on $[t_0, t_1)$ but $u(t_1) = 0$. Since (4.1.3) holds for all $t \in [t_0, t_1)$ it follows by continuity that the bound holds also at t_1 contradicting the hypothesis that $u(t_1) = 0$. A similar argument, involving (4.1.4), is valid in case t_0 is a right-end point of the above interval. The proof is complete.

An interesting uniqueness result and some lower bounds which can be deduced from Theorem 4.1.1 are collected in the next theorem. Here we shall take J to be the nonnegative real line R_+.

THEOREM 4.1.2.   Under the assumption of Theorem 4.1.1 with $J = R_+$

(a)   if $u(t_0) = 0$ for some $t_0 \in R_+$, then $u(t) \equiv 0$ on R_+;

(b)   if $\phi \in L_{2p}[R_+]$ for some p, $1 \leqslant p \leqslant \infty$, then

$$\|u(t)\| \geqslant \|u(t_0\| \exp[-\mu(t-t_0) - C(t-t_0)^{2-(1/p)}], \qquad t \geqslant t_0 \geqslant 0;$$

(c)   if $\phi(t) \leqslant K(t+1)^\alpha$, $\alpha \geqslant 0$, then

$$\|u(t)\| \geqslant \|u(t_0)\| \exp[-\mu(t-t_0) - C(t-t_0)^2(t+1)^{2\alpha}], \qquad t \geqslant t_0 \geqslant 0;$$

(d)   if ϕ is bounded on R_+, then

$$\|u(t)\| \geqslant \|u(t_0)\| \exp[-\mu(t-t_0) - C(t-t_0)^2],$$

where, in (b)–(d), μ is a nonnegative constant depending on the solution u and C is a nonnegative constant depending only on ϕ.

Proof: (a)   Assume that $u(t) \not\equiv 0$. Then $u(t)$ is not identically zero in at least one of the two intervals $[0, t_0)$ and (t_0, ∞). Suppose that $u(t)$ is not identically zero in the first interval $[0, t_0)$. Then, there must exist a subinterval $[t_1, t_2)$ with $0 \leqslant t_1 < t_2 \leqslant t_0$ such that $\|u(t)\| > 0$ for $t_1 \leqslant t < t_2$ and $u(t_2) = 0$. Applying the estimate (4.1.3) with $t = t_2$ and t_0 replaced by t_1, we are lead to a contradiction. Hence $u(t) \equiv 0$ on $[0, t_0]$. Similarly, using the estimate (4.1.4) we obtain a contradiction unless $u(t) \equiv 0$ on $[t_0, \infty)$. This proves the uniqueness part of the theorem.

(b)   Our assumption on ϕ implies by Hölder's inequality

$$\int_{t_0}^t (t-s)\phi^2(s)\,ds \leqslant \left[\int_{t_0}^t [\phi^2(s)\,ds]^p \right]^{(1/p)} \left[\int_{t_0}^t (t-s)^{p/(p-1)}\,ds \right]^{1-(1/p)}$$

$$\leqslant \|\phi\|^2_{L_{2p}[t_0,\infty)}(t-t_0)^{2-(1/p)}. \qquad (4.1.16)$$

Also since ϕ is locally bounded and $\phi \in L_{2p}[R_+]$, we have

$$\int_{t_0}^t \phi(s)\,ds \leqslant \lambda_0(t-t_0), \qquad t_0 \leqslant t \leqslant t_0 + 1, \qquad (4.1.17)$$

where $\lambda_0 = \sup_{t_0 \leqslant t \leqslant t_0+1} \phi(t)$ and

$$\int_{t_0}^t \phi(s)\,ds \leqslant \|\phi\|_{L_{2p}[t_0,\infty)}(t-t_0)^{1-(1/2p)}$$

$$\leqslant \|\phi\|_{L_{2p}[t_0,\infty)}(t-t_0)^{2-(1/p)}, \qquad t > t_0 + 1.$$

This inequality together with (4.1.17), (4.1.16), and the estimate (4.1.3) yields the desired lower bound with $\mu = \lambda_1 + \lambda_0$ and $C = \|\phi\|_{L_{2p}[0,\infty)} + \frac{1}{2}\|\phi\|^2_{L_{2p}[t_0,\infty)}$.

(c)   In this case, one easily gets

$$\frac{1}{2}\int_{t_0}^{t}(t-s)\phi^2(s)\,ds \leqslant \frac{1}{2}K(t-t_0)^2(t+1)^{2\alpha} \qquad (4.1.18)$$

and

$$\int_{t_0}^{t}\phi(s)\,ds \leqslant K(t-t_0)(t+1)^{\alpha} \leqslant K(t-t_0)^2(t+1)^{2\alpha}, \qquad t > t_0 + 1. $$
$$(4.1.19)$$

As before, the inequalities (4.1.18), (4.1.19), (4.1.17), and the estimate (4.1.3) yield the lower bound in (c) with $\mu = \lambda_1 + \lambda_0$ and $C = \frac{3}{2}K$.

(d)   Denoting by C_1 a bound of ϕ on $[t_0, \infty)$ and using (4.1.3) we find

$$\|u(t)\| \geqslant \|u(t_0)\| \exp[-\lambda_1(t-t_0) - C_1(t-t_0) - \tfrac{1}{4}C_1{}^2(t-t_0)^2]$$

and (c) is established with $\mu = \lambda_1 + C_1$ and $C = \frac{1}{4}C_1{}^2$.
[(d) also follows from (c) with $\alpha = 0$.]

COROLLARY 4.1.1.   Under the hypotheses of Theorem 4.1.1 any solution of (4.1.2) is either identically zero on J or never vanishes on J.

PROBLEM 4.1.1.   Utilize (4.1.4) to prove estimates on the solution on R_-.

Let us next derive from Theorem 4.1.1 a global "unique continuation at infinity" result.

THEOREM 4.1.3.   Let the hypotheses of Theorem 4.1.1 be satisfied with $J = R$ and let $\|\phi\|^2_{L_2[R]} = N < \infty$. Assume that for some constant $k > 0$

$$\|u(t)\| = O[\exp(-kt)] \qquad \text{as} \quad t \to -\infty \qquad (4.1.20)$$

and for some $\varepsilon > 0$

$$\|u(t)\| = O[\exp[-(k+N+\varepsilon)t]] \qquad \text{as} \quad t \to +\infty. \qquad (4.1.21)$$

Then $u(t) \equiv 0$ on R.

Proof: Let for some $t_0 \in R$, $\|u(t_0)\| > 0$. Then the estimates (4.1.14) and (4.1.15) are valid and for convenience, we write them in the form

$$\|u(t_0)\| \leqslant \|u(t)\| \exp\left[-Q(t_0)(t-t_0) + \int_{t_0}^t [\phi(s) + \tfrac{1}{2}(t-s)\phi^2(s)]\, ds \right],$$

$$t \geqslant t_0 \qquad\qquad (4.1.22)$$

and

$$\|u(t_0)\| \leqslant \|u(t)\| \exp\left[Q(t_0)(t_0-t) + \int_t^{t_0} [\phi(s) + \tfrac{1}{2}(s-t)\phi^2(s)]\, ds \right],$$

$$t \leqslant t_0. \qquad\qquad (4.1.23)$$

From (4.1.23) and (4.1.20) we obtain, with C standing for a generic constant

$$\|u(t_0)\| \leqslant C\exp[-kt + Q(t_0)(t_0-t) + N^{1/2}(t_0-t)^{1/2} + \tfrac{1}{2}N(t_0-t)]$$
$$\leqslant C\exp[-(k+Q(t_0)+\tfrac{1}{2}N)t + N^{1/2}(t_0-t)^{1/2}], \qquad t \leqslant t_0.$$

Since $\|u(t_0)\| > 0$, we conclude from the last inequality that

$$k + Q(t_0) + \tfrac{1}{2}N \geqslant 0; \qquad\qquad (4.1.24)$$

otherwise as $t \to -\infty$ we get a contradiction.

Similarly, from (4.1.22) and 4.1.21) we obtain

$$\|u(t_0)\| \leqslant C\exp[-(k+N+\varepsilon) - Q(t_0)(t-t_0) + N^{1/2}(t-t_0)^{1/2} + \tfrac{1}{2}N(t-t_0)]$$
$$\leqslant C\exp[-(k+Q(t_0)+\tfrac{1}{2}N+\varepsilon)t + N^{1/2}(t-t_0)^{1/2}], \qquad t \geqslant t_0.$$

Since $\|u(t_0)\| > 0$, we conclude, as before, that

$$k + Q(t_0) + \tfrac{1}{2}N + \varepsilon \leqslant 0$$

which contradicts (4.1.24). The proof is complete.

It is evident from the proof that there is an analogous theorem with the roles of $t = \infty$ and $t = -\infty$ interchanged.

The solutions of (4.1.2) satisfy some convexity-like inequalities which can be used to derive lower bounds for the solutions.

THEOREM 4.1.4. Let the assumptions of Theorem 4.1.1 be satisfied. Let $[a,b]$ be a subinterval of J. Then any solution $u(t)$ of (4.1.2) satisfies the convexity-like property

$$\|u(t)\| \leqslant K\|u(a)\|^{(b-t)/(b-a)}\|u(b)\|^{(t-a)/(b-a)}, \qquad a \leqslant t \leqslant b \quad (4.1.25)$$

where K is a constant depending only on ϕ.

Proof: If $u(a) = 0$, then by Corollary 4.1.1 $u(t) \equiv 0$ and (4.1.25) is trivially true. Assume that $u(a) \neq 0$. From the same corollary it follows that $u(t) \neq 0$

on $[a, b]$. Define

$$m(t) = 2\log\|u(t)\| - \int_a^t v(s)\,ds$$

where $v(s) = 2\,\mathrm{Re}(Lu(t), u(t))/\|u(t)\|^2$. Then, using (4.1.5), we obtain $m''(t) = 2Q'(t)$ and in view of (4.1.11) we get $m''(t) \geqslant -\phi^2(t)$ for $a \leqslant t \leqslant b$. Consider the boundary value problem

$$z''(t) + \phi^2(t) = 0, \qquad z(a) = m(a), \qquad z(b) = m(b). \qquad (4.1.26)$$

It is not difficult to verify that $m(t) \leqslant z(t)$ for $a \leqslant t \leqslant b$, where $z(t)$ is the solution of (4.1.26). Indeed setting $h(t) = m(t) - z(t)$, we note that for $a \leqslant t \leqslant b$

$$h''(t) = m''(t) - z''(t) \geqslant 0 \qquad \text{and} \qquad h(a) = h(b) = 0.$$

Therefore $h(t)$ is concave up on $[a, b]$ and vanishes at the end points of $[a, b]$. This implies that $h(t) \leqslant 0$ for $a \leqslant t \leqslant b$, and our assertion is true.

Let $z_1(t)$ and $z_2(t)$ be the solutions of the boundary value problems

$$z_1'' = 0 \qquad \text{and} \qquad z_1(a) = m(a), \quad z_1(b) = m(b)$$

and

$$z_2'' + \phi^2(t) = 0 \qquad \text{and} \qquad z_2(a) = z_2(b) = 0,$$

respectively. Then $z(t) = z_1(t) + z_2(t)$ where

$$z_1(t) = m(a)(b-t)/(b-a) + m(b)(t-a)/(b-a)$$

and

$$z_2(t) = (b-a)^{-1}\left[(b-t)\int_a^t (s-a)\phi^2(s)\,ds + (t-a)\int_t^b (b-s)\phi^2(s)\,ds\right].$$

Notice that $z_2(t) \leqslant (b-a)\int_a^b \phi^2(s)\,ds$ and $|v(t)| \leqslant 2\phi(t)$. From $m(t) \leqslant z_1(t) + z_2(t)$, it follows, after some manipulation, that

$$\log\|u(t)\| \leqslant [(b-t)/(b-a)]\log\|u(a)\| + [(t-a)/(b-a)]\log\|u(b)\| + C \tag{4.1.27}$$

where

$$C = 2\int_a^b \phi(s)\,ds + [(b-a)/2]\int_a^b \phi^2(s)\,ds.$$

From (4.1.27) the desired inequality (4.1.25) follows with $K = \exp C$. The proof is complete.

In the case of self-adjoint operator A one can use the resolution of the identity associated with A to obtain the following convexity-like statement.

THEOREM 4.1.5. Let A be a self-adjoint operator in H with domain $D(A)$ dense in H. Let $u(t)$ be a solution of (4.1.2) on $[a, b]$ and $Au(t)$ be continuous on $[a, b]$. Assume that $\int_a^b \phi(t)\, dt \leqslant \sqrt{2}/4$. Then

$$\|u(t)\| \leqslant 2\sqrt{2}\, \|u(a)\|^{(b-t)/(b-a)}\, \|u(b)\|^{(t-a)/(b-a)}. \qquad (4.1.28)$$

Proof: Let $\{E_\lambda\}$ be the resolution of the identity associated with A and $E = \int_0^\infty dE_\lambda$ the projection operator in H associated with the positive part of the spectrum of A. Let $u = u(t)$ be the given solution. Set $u_1 = Eu$, $u_2 = (I-E)u$, $f \equiv u' - Au$, $f_1 = Ef$, and $f_2 = (I-E)f$. Then $u_1' - Au_1 = f_1$ and $u_2' - Au_2 = f_2$, so that

$$(d/dt)(u_i, u_i) = 2\operatorname{Re}(Au_i, u_i) + 2\operatorname{Re}(f_i, u_i), \qquad i = 1, 2.$$

Since $(Au_1, u_1) \geqslant 0$ and $(Au_2, u_2) \leqslant 0$, we obtain the inequalities

$$(d/dt)(u_1, u_1) \geqslant 2\operatorname{Re}(f_1, u_1) \qquad \text{and} \qquad (d/dt)(u_2, u_2) \leqslant 2\operatorname{Re}(f_2, u_2). \qquad (4.1.29)$$

Integrating the first inequality in (4.1.29) from t to b and using the fact $|\operatorname{Re}(f_1, u_1)| \leqslant \|f_1\| \|u_1\| \leqslant \|f\| \|u\|$, we get

$$\|u_1(b)\|^2 - \|u_1(t)\|^2 \geqslant 2\operatorname{Re}\int_t^b (f_1(s), u_1(s))\, ds \geqslant -2\int_t^b \|f(s)\| \|u(s)\|\, ds.$$

Setting $M = \max_{a \leqslant t \leqslant b} \|u(t)\|$, there results the inequality

$$\|u_1(t)\|^2 \leqslant \|u_1(b)\|^2 + 2M \int_t^b \|f(s)\|\, ds. \qquad (4.1.30)$$

Similarly, from the second inequality in (4.1.29), we derive

$$\|u_2(t)\|^2 \leqslant \|u_2(a)\|^2 + 2M \int_a^t \|f(s)\|\, ds. \qquad (4.1.31)$$

Adding (4.1.30) and (4.1.31) and remembering that u_1 and u_2 are orthogonal vectors, one readily sees that

$$\|u(t)\|^2 \leqslant \|u_2(a)\|^2 + \|u_1(b)\|^2 + 2M \int_a^b \|f(s)\|\, ds. \qquad (4.1.32)$$

Taking the maximum in both sides of (4.1.32) and using the inequality $2MN \leqslant (M/\sqrt{2})^2 + (\sqrt{2}N)^2$, it follows that

$$M^2 \leqslant \|u_2(a)\|^2 + \|u_1(b)\|^2 + (M^2)/2 + 2\left(\int_a^b \|f(s)\| \, ds\right)^2.$$

Therefore

$$\max_{a \leqslant t \leqslant b} \|u(t)\|^2 \leqslant 2(\|u(a)\|^2 + \|u(b)\|^2) + 4\left(\int_a^b \|u'(s) - Au(s)\| \, ds\right)^2.$$

$$(4.1.33)$$

We now set $w = \exp(\sigma t)u(t)$ with σ real and notice that

$$\|w' - (A + \sigma I)w\| \leqslant \exp(\sigma t)\|u' - Au\|.$$

Applying (4.1.33) with A replaced by $A + \sigma I$, using (4.1.2), and the assumption on ϕ, we obtain

$$\max_{a \leqslant t \leqslant b} \|\exp(\sigma t)u(t)\|^2 \leqslant 2[\|\exp(\sigma a)u(a)\|^2 + \|\exp(\sigma b)u(b)\|^2]$$

$$+ 4\left(\int_a^b \|\exp(\sigma s)\phi(s)u(s)\| \, ds\right)^2$$

$$\leqslant 2[\|\exp(\sigma a)u(a)\|^2 + \|\exp(\sigma b)u(b)\|^2]$$

$$+ \tfrac{1}{2} \max_{a \leqslant t \leqslant b} \|\exp(\sigma t)u(t)\|.$$

Hence

$$\|\exp(\sigma t)u(t)\|^2 \leqslant 4\|\exp(\sigma a)u(a)\|^2 + 4\|\exp(\sigma b)u(b)\|^2.$$

Choose σ so that the two terms on the right become equal, i.e.,

$$\sigma = (b-a)^{-1}\log(\|u(a)\|/\|u(b)\|)$$

and the desired inequality (4.1.28) follows. The proof is complete.

Using the convexity-like inequality (4.1.28) one can obtain lower bounds for the solution $u(t)$. This we state as

PROBLEM 4.1.2. Assume that $\phi(t)$ is integrable on every finite interval of R_+. Starting with $t_0 = 0$ let t_n with $n = 1, 2, \ldots$ be such that

$$\int_{t_{n-1}}^{t_n} \phi(t) \, dt = \sqrt{2}/12$$

and set $t_{n+1} - t_n = \rho_n$. If there are only a finite number of such intervals, the last has infinite length and the integral of ϕ over it does not exceed $\sqrt{2}/12$. Suppose that for some numbers C and K

$$\sum_{j=0}^{n} \rho_j^{-1} \leqslant K \left(1 + \sum_{j=0}^{n} \rho_j\right)^C, \qquad n = 1, 2, \ldots$$

Then any solution $u(t)$ of (4.1.2) with A self-adjoint, satisfies

$$\|u(t)\| \geqslant \|u(0)\| \exp[-\mu(t+1)^{C+1}] \beta^t, \qquad t \geqslant t_2$$

where μ is a fixed constant, while β is a constant depending on the solution. In particular

(i) if $\phi \in L_p(R_+)$ and $1 \leqslant p \leqslant 2$, then

$$\|u(t)\| \geqslant \|u(0)\| \exp\{-\mu(t+1)\} \beta^t, \qquad t \geqslant t_2;$$

(ii) if $\phi \in L_p(R_+)$ and $2 < p \leqslant \infty$, then

$$\|u(t)\| \geqslant \|u(0) \exp\{-\mu(t+1)^{2-(2/p)}\} \beta^t, \qquad t \geqslant t_2;$$

(iii) if $\phi(t) \leqslant K(1+t)^C$ for $C \geqslant 0$, then

$$\|u(t)\| \geqslant \|u(0)\| \exp\{-\mu(t+1)^{2C+2}\} \beta^t, \qquad t \geqslant t_2.$$

[Hint: Apply (4.1.28) with $t = t_j$, $a = t_{j-1}$, and $b = t_{j+1}$. Set $\sigma_j = \|u(t_j)\|$ and estimate $(\sigma_j - \sigma_{j-1})/\rho_{j-1}$.]

It is clear from Theorem 4.1.1 and the subsequent considerations that finding an estimate for the function $Q(t)$ is indeed essential. This has been achieved in the case of a symmetric operator A, as the foregoing discussion shows. We give below sufficient conditions which guarantee a similar estimate when A is not symmetric.

THEOREM 4.1.6. Assume that for λ sufficiently large, $\lambda \in \rho(A)$ and the following conditions hold:

(i) $$\lim_{\lambda \to \infty} \lambda R(\lambda; A) x = x, \qquad x \in H;$$

(ii) $$\|\lambda R(\lambda; A) x\|^2 \geqslant [1 - 2K/\lambda] \|x\|^2, \qquad \lambda \text{ sufficiently large.}$$

Then every solution $u(t)$ of (4.1.2) satisfies the inequality

$$\|u(t)\| \geqslant \|u(t_0)\| \exp\left[-K(t-t_0) - \int_{t_0}^{t} \phi(s) \, ds\right]. \qquad (4.1.34)$$

Proof: Set $m(t) = \|u(t)\|^2$. Then

$$m'(t) = 2\,\mathrm{Re}(u'(t), u(t))$$

$$= 2\,\mathrm{Re}(Au(t), u(t)) + 2\,\mathrm{Re}(Lu(t), u(t)). \qquad (4.1.35)$$

Observe that

$$\lambda R(\lambda; A)u = u + \lambda^{-1}Au + \lambda^{-1}[\lambda R(\lambda, A)Au - Au]$$

$$= u + \lambda^{-1}Au + \lambda^{-1}\tilde{R}(\lambda)$$

where by (i) we have $\tilde{R}(\lambda) \equiv \lambda R(\lambda, A)Au - Au \to 0$ as $\lambda \to \infty$.
We thus have

$$\|\lambda R(\lambda, A)u\|^2 = \|u\|^2 + \lambda^{-1}2\,\mathrm{Re}(Au, u) + \lambda^{-1}2\,\mathrm{Re}(u, \tilde{R}(\lambda)) + \lambda^{-2}\|Au\|^2$$

$$+ \lambda^{-2}2\,\mathrm{Re}(Au, \tilde{R}(\lambda)) + \lambda^{-2}\|\tilde{R}(\lambda)\|^2.$$

From this identity and (ii) we obtain

$$-2K\|u\|^2 \leqslant 2\,\mathrm{Re}(Au, u) + 2\,\mathrm{Re}(u, \tilde{R}(\lambda)) + \lambda^{-1}\|Au\|^2$$

$$+ \lambda^{-1}2\,\mathrm{Re}(Au, \tilde{R}(\lambda)) + \lambda^{-1}\|\tilde{R}(\lambda)\|^2. \qquad (4.1.36)$$

Since $\tilde{R}(\lambda) \to 0$ as $\lambda \to \infty$, it follows from (4.1.36) that

$$\mathrm{Re}(Au(t), u(t)) \geqslant -Km(t).$$

This together with (4.1.35) and (4.1.2) implies that

$$m'(t) \geqslant -2[K + \phi(t)]m(t)$$

which yields the desired inequality (4.1.34). The proof is complete.

We should remark that hypothesis (i) is satisfied if, for example, A generates a strongly continuous semigroup in H.

4.2. Lower Bounds, Uniqueness, and Convexity (General Results)

Consider the time dependent evolution operator

$$Lu = u' - A(t)u, \qquad ' = d/dt \qquad (4.2.1)$$

in a Hilbert space H. We assume that $A(t)$ admits a decomposition of the form

$$A(t) = A_+(t) + \delta(A_-'(t) + A_-''(t)), \qquad \delta \geqslant 0 \qquad (4.2.2)$$

where $A_+(t)$ is a linear symmetric operator over $D[A(t)]$ while $A_-'(t)$ and $A_-''(t)$ are linear skew symmetric over $D[A(t)]$.

Here we shall obtain results analogous to those of Section 4.1 for the general evolution inequality

$$\|Lu(t)\| \leqslant \Phi\ (t, m(t), q(t)), \qquad t \in J \tag{4.2.3}$$

where J is an interval on the real line R, $A(t)$ admits the decomposition (4.2.2), $m(t) = \|u(t)\|^2 + P[u(t)]$ with $P[u(t)]$ a nonnegative, nonlinear functional defined on $D[A(t)]$, $q(t) = (A_+(t)u(t), u(t))$, and $\Phi: J \times R_+ \times R \to R_+$.

The form of the inequality (4.2.3) will offer a wide range of applicability of these results to partial differential equations, but we shall be mainly concerned here with abstract inequalities.

DEFINITION 4.2.1. A function $u \in C[J, H]$ is said to be a solution of the inequality (4.2.3) if

   (i) $u(t) \in D[A(t)]$ for $t \in J$ and $A(t)u(t)$ is continuous for $t \in J$;

   (ii) $du(t)/dt$ exists and is continuous on J;

   (iii) $u(t)$ satisfies the inequality (4.2.3) for $t \in J$.

We shall first obtain general lower bounds and two uniqueness theorems for the solutions of (4.2.3) on the interval J of the real line R. We shall often use the inequality $m(t) \geqslant \|u(t)\|^2$ which is valid since $P[u(t)] \geqslant 0$. We also define $Q(t) = q(t)/m(t)$ as long as $m(t) \neq 0$. For easy references we state the following hypotheses:

Hypothesis 1: There exists a function $\phi \in C[J \times R, R_+]$ such that for all $t \in J$ with $m(t) \neq 0$

$$\Phi[t, m(t), q(t)] \leqslant m(t)^{1/2} \phi[t, Q(t)].$$

Hypothesis 2: The functional $P[u(t)]$ is differentiable with respect to t and for all $t \in J$

$$|(d/dt) P[u(t)]| \leqslant 2\omega(t) m(t)$$

where $\omega \in C[J, R_+]$.

Hypothesis 3: There exists functions $\psi_i \in C[J \times R, R]$ with $i = 1, 2, 3$ such that, for any solution $u(t)$ of (4.2.3) with $m(t) \neq 0$ and for any number

δ such that $0 \leqslant \delta < \frac{2}{3}$, the following estimates hold:

(i) $\delta \operatorname{Re}(A_+(t)u(t), A_-'(t)u(t))$

$$\geqslant -\delta m(t)\psi_1(t, Q(t)) - \delta \|A_+(t)u(t) - Q(t)u(t)\|^2;$$

(ii) $\delta \|A_-''(t)u(t)\|^2 \leqslant \delta m(t)\psi_2(t, Q(t)) + \delta \|A_+(t)u(t) - Q(t)u(t)\|^2;$

(iii) the function $(A_+(t)u(t), u(t))$ is differentiable on J and

$$(d/dt)(A_+(t)u(t), u(t)) - 2\operatorname{Re}(A(t)u(t), u'(t))$$

$$\geqslant -m(t)\psi_3(t, Q(t)) - \delta \|A_+(t)u(t) - Q(t)u(t)\|^2.$$

It should be noticed that when $A(t)$ is symmetric then $\delta = 0$ and Hypotheses 3(i) and 3(ii) are automatically satisfied.

Hypothesis 4:   There exists a function $\theta_1 \in C[J \times R, R]$ such that $\theta_1(t, y)$ is nondecreasing in y for each $t \in J$ and

$$2[y - \phi(t, y) - \omega(t)] \geqslant \theta_1(t, y), \qquad t \geqslant t_0, \quad t \in J.$$

Hypothesis 5:   There exists a function $\theta_2 \in C[J \times R, R]$ such that $\theta_2(t, y)$ is nondecreasing in y for each $t \in J$ and

$$2[y + \phi(t, y) + \omega(t)] \leqslant \theta_2(t, y), \qquad t \leqslant t_0, \quad t \in J.$$

We define the function

$$w(t, y) \equiv \psi(t, y) + 2\omega(t)|y| + (2 - 5\delta)^{-1}\phi^2(t, y) \qquad (4.2.4)$$

where $\psi(t, y) \equiv 2\delta\psi_1(t, y) + \delta\psi_2(t, y) + \psi_3(t, y)$. We also denote by $y_1(t)$ and $y_2(t)$ the right minimal and the left maximal solution, respectively, of the scalar initial value problem

$$y' = -w(t, y) \qquad \text{and} \qquad y(t_0) = Q(t_0). \qquad (4.2.5)$$

We now pass to our main result of this section.

THEOREM 4.2.1.   Let $u(t)$ be a solution of (4.2.3) and let t_0 be a point in J. Then

(a)   under Hypotheses 1–4 the following lower bound is valid:

$$m(t) \geqslant m(t_0)\exp \int_{t_0}^t \theta_1(s, y_1(s))\, ds, \qquad t \geqslant t_0, \quad t \in J; \qquad (4.2.6)$$

(b) under Hypotheses 1–3 and 5 the following lower bound is valid:

$$m(t) \geqslant m(t_0) \exp \int_t^{t_0} -\theta_2(s, y_2(s))\, ds, \qquad t \leqslant t_0, \quad t \in J. \quad (4.2.7)$$

Proof: Suppose first that $m(t) > 0$ for all $t \in J$. Then, using the decomposition (4.2.2) the symmetry of A_+ and the skew symmetry of A_-' and A_-'', we obtain

$$(d/dt)\, m(t) = 2\,\mathrm{Re}(u'(t), u(t)) + (d/dt)\, P(u(t))$$
$$= 2Q(t)\, m(t) + 2\mathrm{Re}(Lu(t), u(t)) + (d/dt)\, P(u(t)). \quad (4.2.8)$$

In view of Hypothesis 1 and (4.2.3) we have from (4.2.8)

$$|(d/dt)\, m(t) - 2Q(t)\, m(t) - (d/dt)\, P(u(t))| \leqslant 2\Phi(t, m(t), q(t))\, \|u(t)\|$$
$$\leqslant 2\phi(t, Q(t))\, m(t). \quad (4.2.9)$$

By (4.2.9) and Hypothesis 2 we are led to the inequalities

$$(d/dt)\, m(t) \geqslant 2[Q(t) - \phi(t, Q(t)) - \omega(t)]\, m(t), \qquad t \in J, \quad (4.2.10)$$

and

$$(d/dt)\, m(t) \leqslant 2[Q(t) + \phi(t, Q(t)) + \omega(t)]\, m(t), \qquad t \in J. \quad (4.2.11)$$

Since $A_+(t) u(t)$ is continuous in t and $A_+(t)$ is symmetric, it is easily seen, by taking the difference quotient and passing to the limit, that the function $(A_+(t) u(t), u(t))$ is differentiable and

$$(d/dt)(A_+(t) u(t), u(t)) = (\dot{A}_+(t) u(t), u(t)) + 2\,\mathrm{Re}(A_+(t) u(t), u'(t)) \quad (4.2.12)$$

where

$$(\dot{A}_+(t) u(t), u(t)) \equiv (d/dt)(A_+(t) u(t), u(t)) - 2\,\mathrm{Re}(A_+(t) u(t), u'(t)).$$

It follows that (suppressing the variable t) and using (4.2.2)

$$dQ/dt = (\dot{A}_+ u, u)/m - (Q/m)(d/dt)\, P(u)$$
$$+ (2/m)[\|A_+ u\|^2 - Q^2 m + \mathrm{Re}(A_+ u - Qu, Lu)]$$
$$+ (2\delta/m)\,\mathrm{Re}(A_+ u, A_-' u) + (2\delta/m)\,\mathrm{Re}(A_+ u, A_-'' u)$$
$$\equiv I_1 + I_2 + I_3 + I_4 + I_5. \quad (4.2.13)$$

We shall estimate the terms I_i with $i = 1, \ldots, 5$.

From Hypothesis 3(iii) we get

$$I_1 \geqslant -\psi_3(t, Q) - (\delta/m)\|A_+u - Qu\|^2. \tag{4.2.14}$$

By Hypothesis 2 we see that

$$I_2 \geqslant -2|Q|\omega(t). \tag{4.2.15}$$

Notice that

$$\begin{aligned}
\|A_+u - Qu\|^2 &= \|A_+u\|^2 - 2Q(A_+u, u) + Q^2\|u\|^2 \\
&\leqslant \|A_+u\|^2 - 2Q^2m + Q^2m \\
&= \|A_+u\|^2 - Q^2m.
\end{aligned} \tag{4.2.16}$$

Next, using the arithmetic-geometric mean inequality on $2\,\mathrm{Re}(A_+u - Qu, Lu)$, we obtain, for any $a > 0$

$$\begin{aligned}
2\,\mathrm{Re}(A_+u - Qu, Lu) &\geqslant -2\|A_+u - Qu\|\,\|Lu\| \\
&\geqslant -a\|A_+u - Qu\|^2 - a^{-1}\|Lu\|^2 \\
&\geqslant -a\|A_+u - Qu\|^2 - (m/a)\phi^2(t, Q). \tag{4.2.17}
\end{aligned}$$

The estimates (4.2.16) and (4.2.17) yield

$$I_3 \geqslant (2-a)/m\,\|A_+u - Qu\|^2 - a^{-1}\phi^2(t, Q). \tag{4.2.18}$$

In view of Hypothesis 3(i), it follows that

$$I_4 \geqslant -2\delta\psi_1(t, Q) - (2\delta/m)\|A_+u - Qu\|^2. \tag{4.2.19}$$

By Hypothesis 3(ii) and the skew symmetry of A''_- one gets

$$\begin{aligned}
\delta|2\,\mathrm{Re}(A_+u, A''_-u)| &= \delta|2\,\mathrm{Re}(A_+u - Qu, A''_-u)| \\
&\leqslant \delta\|A_+u - Qu\|^2 + \delta\|A''_-u\| \\
&\leqslant 2\delta\|A_+u - Qu\|^2 + \delta m\psi_2(t, Q)
\end{aligned}$$

which implies that

$$I_5 \geqslant -(2\delta/m)\|A_+u - Qu\|^2 - \delta\psi_2(t, Q). \tag{4.2.20}$$

Using (4.2.14), (4.2.15), (4.2.18), (4.2.19), and (4.2.20) in (4.2.13) and choosing $a = 2 - 5\delta$, we finally obtain

$$dQ/dt \geqslant -(2\delta\psi_1 + \delta\psi_2 + \psi_3) - 2|Q|\omega - (2 - 5\delta)^{-1}\phi^2(t, Q),$$

that is, recalling (4.2.4)

$$dQ/dt \geqslant -w(t, Q). \tag{4.2.21}$$

The differential inequality (4.2.21) now yields [42]

$$Q(t) \geqslant y_1(t), \qquad t \geqslant t_0, \quad t \in J, \tag{4.2.22}$$

and

$$Q(t) \leqslant y_2(t), \qquad t \leqslant t_0, \quad t \in J. \tag{4.2.23}$$

Now assume that Hypothesis 4 is satisfied. Then, using (4.2.10) and (4.2.22) we get

$$(d/dt)m(t) \geqslant \theta_1(t, y_1(t))m(t), \qquad t \geqslant t_0, \quad t \in J,$$

from which the estimate (4.2.6) follows and (a) is proved. Similarly, assuming that Hypothesis 5 is satisfied, we derive from (4.2.11) and (4.2.23) the inequality

$$(d/dt)m(t) \leqslant \theta_2(t, y_2(t))m(t), \qquad t \leqslant t_0, \quad t \in J,$$

from which the estimate (4.2.7) follows and (b) is proved.

We have established the desired lower bounds under the additional condition that $m(t) > 0$ for $t \in J$. We shall now remove this assumption. If $m(t_0) = 0$ the estimates (4.2.6) and (4.2.7) are clearly valid. Now assume that $m(t_0) > 0$. We shall prove that $m(t) > 0$ for all $t \in J$ and hence the previous arguments are valid. Otherwise there exists an interval with one end point t_0, say, $[t_0, t_1)$, such that $m(t) > 0$ on $[t_0, t_1)$ but $m(t_1) = 0$. Since (4.2.6) holds for all $t \in [t_0, t_1)$ it follows by continuity that the same bound holds also at $t = t_1$ contradicting the hypothesis $m(t_1) = 0$. A similar argument, involving (4.2.7) is valid in case t_0 is a right-end point of the above interval. The proof is therefore complete.

A consequence of Theorem 4.2.1. is the following interesting uniqueness result.

THEOREM 4.2.2. Under Hypotheses 1–5 for any solution $u(t)$ of (4.2.3) either $m(t) > 0$ for all $t \in J$ or $m(t) \equiv 0$ on J. In the special case when $P[u(t)] \equiv 0$, if $u(t_0) = 0$ for some $t_0 \in J$, then $u(t) \equiv 0$ on J.

Proof: It is clear from (4.2.6) and (4.2.7) that if $m(t_0) > 0$ then $m(t) > 0$ for all $t \in J$. Next, assume that $m(t_0) = 0$ for some $t_0 \in J$. We shall prove that $m(t) \equiv 0$ for all $t \in J$. If not, then $m(t)$ is not identically zero in an interval either to the left or to the right of t_0. Suppose that this happens to the left of t_0. Then, there must exist a subinterval $[t_1, t_2]$ of J such that $m(t) > 0$ for $t_1 \leqslant t < t_2$ and $m(t_2) = 0$. Applying the estimate (4.2.6) with

$t = t_2$ and $t_0 = t_1$ we obtain a contradiction. Hence $m(t) \equiv 0$ to the left of t_0. Similarly, using the estimate (4.2.7) we obtain a contradiction unless $u(t) \equiv 0$ to the right of t_0. The proof is complete.

The following (unique continuation at infinity) theorem shows that the solutions of (4.2.3) (actually $m(t)$) cannot tend to zero too rapidly as $t \to \infty$ unless they are identically zero.

THEOREM 4.2.3. Let Hypotheses 1–5 be satisfied on the whole real line R and $u(t)$ be a solution of (4.2.3). Assume that there exist constants k, l, N such that $k \geqslant 0$, $l > 0$, and N depends on the solution satisfying the order relations

$$m(t) = O[\exp(-kt)] \qquad \text{as} \quad t \to -\infty, \tag{4.2.24}$$

$$m(t) = O[\exp[-(k+l)t]] \qquad \text{as} \quad t \to +\infty, \tag{4.2.25}$$

$$\exp \int_{t_0}^{t} -\theta_1[s, y_1(s)]\, ds = O[\exp(-Nt)] \qquad \text{as} \quad t \to +\infty, \tag{4.2.26}$$

and

$$\exp \int_{t}^{t_0} \theta_2[s, y_2(s)]\, ds = O[\exp(-Nt)] \qquad \text{as} \quad t \to -\infty. \tag{4.2.27}$$

Then $m(t) \equiv 0$.

Proof: Let $m(t_0) > 0$ for some t_0. Then the estimates (4.2.6) and (4.2.7) are valid and for convenience we write them in the form

$$m(t_0) \leqslant m(t)\exp \int_{t_0}^{t} -\theta_1[s, y_1(s)]\, ds, \qquad t \geqslant t_0 \tag{4.2.28}$$

and

$$m(t_0) \leqslant m(t)\exp \int_{t}^{t_0} \theta_2[s, y_2(s)]\, ds, \qquad t \leqslant t_0. \tag{4.2.29}$$

From (4.2.28), (4.2.25), and (4.2.26) we obtain, with C standing for a generic constant, the inequality

$$m(t_0) \leqslant C\exp[-(k+l+N)t] \qquad \text{as} \quad t \to +\infty.$$

Since $m(t_0) > 0$, it is necessary that

$$k + l + N \leqslant 0. \tag{4.2.30}$$

Similarly from (4.2.29), (4.2.24), and (4.2.27) we get

$$m(t_0) \leqslant C \exp[-(k+N)t] \qquad \text{as} \quad t \to -\infty$$

and therefore $k + N \geqslant 0$ which contradicts (4.2.30) and the proof is complete. It is evident from the proof that there is an analogous theorem with the roles of $t = \infty$ and $t = -\infty$ interchanged.

As an example we shall study the evolution inequality

$$\|du/dt - A(t)u(t)\| \leqslant [\phi_1(t)m(t) + \phi_2(t)a_t(u(t), u(t))]^{\frac{1}{2}},$$

$$t \in R \tag{4.2.31}$$

where $A(t)$ admits the decomposition (4.2.2) in the Hilbert space H

$$m(t) = \|u(t)\|^2 + \int_t^T 2\omega(s)\|u(s)\|^2 \, ds,$$

a_t is a symmetric positive semidefinite, bilinear functional defined on $D[A(t)]$, ϕ_1 and ϕ_2 are nonnegative functions in $L_1(R)$, and $\omega(t)$ a nonnegative continuous function on R. In addition we shall make the following assumptions:

There exist nonnegative measurable functions α, γ_i, β_i, and α_i with $i = 1, 2, 3$ bounded on every closed finite subinterval of R such that for any solution $u(t)$ of (4.2.31) with $m(t) \neq 0$ and for any number δ where $0 \leqslant \delta < 2/5$ we have

Assumption 1:

$$\delta \operatorname{Re}(A_+ u, A_-'u) \geqslant -\delta\gamma_1 \|A_+ u\| \|u\| - \delta\beta_1 \|u\|^2 - \delta\alpha_1 a_t(u, u).$$

Assumption 2:

$$\delta \|A_-''u\|^2 \leqslant \delta\gamma_2 \|A_+ u\| \|u\| + \delta\beta_2 \|u\|^2 + \delta\alpha_2 a_t(u, u).$$

Assumption 3: The function $(A_+(t)u(t), u(t))$ is differentiable on J and

$$(\dot{A}_+ u, u) \geqslant -\delta\gamma_3 \|A_+ u\| \|u\| - \beta_3 \|u\|^2 - \alpha_3 a_t(u, u).$$

Assumption 4: If $\phi_2 + \alpha_1 + \alpha_2 + \alpha_3 \neq 0$, we shall assume that

$$-(A_+ u, u) \geqslant a_t(u, u) - \alpha \|u\|^2.$$

First, our aim is to show that Hypotheses 1–5 of Theorem 4.2.1 are also true for (4.2.31). Then we shall obtain explicit lower bounds for the solutions of (4.2.31).

If $\phi_2(t) \not\equiv 0$ from Assumption 4 we get (suppressing t and using the same notation for q and Q as before)

$$[\phi_1 m + \phi_2 a_t(u,u)]^{\frac{1}{2}} \leqslant [\phi_1 m + \phi_2 \alpha m - \phi_2 q]^{\frac{1}{2}}$$
$$\leqslant m^{\frac{1}{2}}[\phi_1 + \phi_2(\alpha - Q)]^{\frac{1}{2}}, \qquad m \neq 0.$$

Hence Hypothesis 1 is satisfied with

$$\phi[t,Q] = [\phi_1 + \phi_2(\alpha - Q)]^{\frac{1}{2}}. \tag{4.2.32}$$

Notice also that because of Assumption 4 $\alpha - Q \geqslant 0$. If $\phi_2(t) \equiv 0$, then $\phi(t,Q) = \phi_1^{\frac{1}{2}}$. Hence in any case $\phi(t,Q)$ is given by (4.2.32). Clearly Hypothesis 2 is satisfied with

$$P[u(t)] = \int_t^T 2\omega(s)\|u(s)\|^2 \, ds.$$

The proof that Hypothesis 3 is satisfied requires a trick.

Let θ denote the angle between the vectors $A_+ u$ and u in the Hilbert space H. Then

$$\|A_+ u - Qu\|^2 \geqslant \|A_+ u\|^2 \sin^2 \theta$$

and

$$|Q| = m^{-1}|(A_+ u, u)| = m^{-1}\|A_+ u\| \|u\| |\cos \theta|.$$

From these inequalities we obtain, using the δ of Assumptions 1–4

$$\delta\gamma_i \|A_+ u\| \|u\| = \delta\gamma_i \|A_+ u\| \|u\| (\sin^2 \theta + \cos^2 \theta)$$
$$\leqslant \delta\gamma_i \|A_+ u\| \|u\| \sin^2 \theta + \delta\gamma_i \|A_+ u\| \|u\| |\cos \theta|$$
$$\leqslant \delta(\|A_+ u\|^2 + \gamma_i^2 \|u\|^2)\sin^2 \theta + \delta\gamma_i |Q| m$$
$$\leqslant \delta\|A_+ u - Qu\|^2 + \delta(\gamma_i^2 + \gamma_i |Q|)m, \qquad i = 1, 2, 3.$$
$$\tag{4.2.33}$$

In view of Assumptions 1–4 and (4.2.33) we get, as long as $m(t) \neq 0$

(i) $\quad \delta \operatorname{Re}(A_+ u, A_-'u) \geqslant -\delta \|A_+ u - Qu\|^2$
$$\qquad\qquad - \delta m(\gamma_1^2 + \gamma_1 |Q| + \beta_1 - \alpha_1 Q + \alpha\alpha_1);$$

(ii) $\quad \delta \|A''_- u\|^2 \leqslant \delta \|A_+ u - Qu\|^2 + \delta m(\gamma_2^2 + \gamma_2 |Q| + \beta_2 - \alpha_2 Q + \alpha\alpha_2);$

(iii) $\quad (\dot{A}_+ u, u) \geqslant -\delta \|A_+ u - Qu\|^2 - \delta m(\gamma_3^2 + \gamma_3 |Q| + \beta_3 + \alpha_3 Q - \alpha\alpha_3).$

Therefore Hypothesis 3 is established with

$$\psi_i(t, Q) = \gamma_i^2 + \gamma_i |Q| + \beta_i - \alpha_i Q + \alpha\alpha_i, \qquad i = 1, 2, 3,$$

and

$$\psi(t, Q) = -(2\delta\alpha_1 + \delta\alpha_2 + \alpha_3) Q + (2\delta\gamma_1 + \delta\gamma_2 + \gamma_3)|Q|$$
$$+ (2\delta\gamma_1^2 + \delta\gamma_2^2 + \gamma_3^2) + (2\delta\beta_1 + \delta\beta_2 + \beta_3) + \alpha(2\delta\alpha_1 + \delta\alpha_2 + \alpha_3).$$

$$(4.2.34)$$

Next we shall verify Hypotheses 4 and 5. Since the function $y - \phi(t, y) = y - [\phi_1 + \phi_2(\alpha - y)]^{1/2}$ with $y \leqslant \alpha$ is increasing in y we could take for $\theta_1(t, y)$ the function $2[y - \phi(t, y) - \omega(t)]$. However, a simpler function can be chosen as follows:

$$y - [\phi_1 + \phi_2(\alpha - y)]^{1/2} \geqslant y - \phi_1^{1/2} - [\phi_2(\alpha - y)]^{1/2}$$
$$\geqslant y - \phi_1^{1/2} - (\phi_2/2\rho) - \rho(\alpha - y)$$
$$= (1 + \rho)y - \phi_1^{1/2} - (\phi_2/2\rho) - \alpha\rho, \qquad \rho > 0.$$

Therefore one can take

$$\theta_1(t, y) \equiv 2[(1 + \rho)y - \phi_1^{1/2} - (\phi_2/2\rho) - \alpha\rho - \omega],$$

and Hypothesis 4 is satisfied.

Similarly

$$y + [\phi_1 + \phi_2(\alpha - y)]^{1/2} \leqslant y + \phi_1^{1/2} + [\phi_2(\alpha - y)]^{1/2}$$
$$\leqslant y + \phi_1^{1/2} + \phi_2/2\rho + \rho(\alpha - y)$$
$$= (1 - \rho)y + \phi_1^{1/2} + \phi_2/2\rho + \alpha\rho.$$

So for any $0 < \rho < 1$ we may take

$$\theta_2(t, y) \equiv 2[(1 - \rho)y + \phi_1^{1/2} + \phi_2/2\rho + \alpha\rho + \omega]$$

and Hypothesis 5 is satisfied.

From (4.2.4), (4.2.34), and (4.2.32) we get

$$w(t, y) = -A(t)y + B(t)|y| + C(t) \qquad (4.2.35)$$

where

$$A(t) = 2\delta\alpha_1 + \delta\alpha_2 + \alpha_3 + (2 - 5\delta)^{-1}\phi_2$$

$$B(t) = 2\delta\gamma_1 + \delta\gamma_2 + \gamma_3 + 2\omega$$

$$C(t) = (2\delta\gamma_1^2 + \delta\gamma_2^2 + \gamma_3^2) + (2\delta\beta_1 + \delta\beta_2 + \beta_3) + \alpha(2\delta\alpha_1 + \delta\alpha_2 + \alpha_3)$$
$$+ (2 - 5\delta)^{-1}(\phi_1 + \alpha\phi_2).$$

Finally we must compute the solutions of (4.2.5). We first make the observation that the solution of the equation $y' = Ay - B|y| - C$ with $y(t_0) = y_0$ is given by the formula

$$y(t) = y_0 \exp \int_{t_0}^{t} (A \pm B) \, ds - \int_{t_0}^{t} \left[C(s) \exp \int_{s}^{t} (A \pm B) \, d\xi \right] ds$$

and therefore if $y_0 \leqslant 0$ then $y(t) \leqslant 0$ for $t \geqslant t_0$ and if $y_0 \geqslant 0$, then $y(t) \geqslant 0$ for $t \leqslant t_0$.

Set

$$\lambda_1 = \min(0, Q(t_0)) \qquad \text{and} \qquad \lambda_2 = \max(0, Q(t_0)).$$

By what we have said above the solution $y_1^*(t)$ of the equation

$$y' = Ay - B|y| - C \qquad \text{and} \qquad y(t_0) = \lambda_1, \qquad t \geqslant t_0$$

is negative for $t \geqslant t_0$ and therefore is given by

$$y_1^*(t) = \lambda_1 \exp \int_{t_0}^{t} (A + B) \, ds - \int_{t_0}^{t} \left[C(s) \exp \int_{s}^{t} (A + B) \, d\xi \right] ds, \qquad t \geqslant t_0.$$

From the theory of differential inequalities it follows that $y_1(t) \geqslant y_1^*(t)$ and the estimate (4.2.6) takes an explicit form with $y_1(t)$ replaced by $y_1^*(t)$. Similarly, if $y_2^*(t)$ is the solution of the equation

$$y' = Ay - B|y| - C \qquad \text{and} \qquad y(t_0) = \lambda_2, \qquad t \leqslant t_0,$$

then $y_2^*(t)$ is positive for $t \leqslant t_0$ and is given by

$$y_2^*(t) = \lambda_2 \exp \int_{t_0}^{t} (A - B) \, ds - \int_{t_0}^{t} \left[C(s) \exp \int_{s}^{t} (A - B) \, d\xi \right] ds, \qquad t \leqslant t_0.$$

Also $y_2^*(t) \geqslant y_2(t)$ and the estimate (4.2.7) takes an explicit form with $y_2(t)$ replaced by $y_2^*(t)$. The proof is complete.

The following two corollaries are with respect to the inequalities

$$\|du/dt - A(t)u(t)\|^2 \leqslant \phi_1(t)\|u(t)\|^2 + \phi_2(t) a_t(u(t), u(t)),$$

$$t \in R \tag{4.2.36}$$

and

$$\|du/dt - A(t)u(t)\|^2 \leqslant \phi(t) \left[\|u(t)\|^2 + \int_{t}^{T} \omega(s)\|u(s)\|^2 \, ds \right],$$

$$t \in [0, T) \tag{4.2.37}$$

COROLLARY 4.2.1.   When $\delta = 0$ (that is, when $A = A_+$), $\omega(t) \equiv 0$, $\gamma_i(t) \equiv 0$, $i = 1, 2, 3$, $\beta_i(t) \equiv \alpha_i(t) \equiv 0$, $i = 1, 2$, and under Assumptions 3 and 4 we have

$$dQ/dt \geqslant (\alpha_3 + \tfrac{1}{2}\phi_2) Q - (\beta_3 + \alpha\alpha_3 + \tfrac{1}{2}\phi_1 + \tfrac{1}{2}\alpha\phi_2). \qquad (4.2.38)$$

This inequality follows directly from (4.2.35) and (4.2.21).

COROLLARY 4.2.2.   When $\phi_2 = 0$, $\delta \neq 0$ (for example, $\delta = 1/5$), $\alpha = \alpha_1 = \alpha_2 = \alpha_3 \equiv 0$, $\phi_1 = \phi$, and under the Assumptions 1–3

$$\ddot{l}(t) + B(t)|\dot{l}(t)| + 2C(t) = 0 \qquad (4.2.39)$$

where $l(t) = 2Q(t)$, $B(t) = \tfrac{2}{5}\gamma_1 + \tfrac{1}{5}\gamma_2 + \gamma_3 + 2\omega$, and $C(t) = \tfrac{2}{5}\gamma_1{}^2 + \tfrac{1}{5}\gamma_2{}^2 + \gamma_3)$ $+ (\tfrac{2}{5}\beta_1 + \tfrac{1}{5}\beta_2 + \beta_3) + \phi^2$.
  This inequality follows directly from (4.2.35) and (4.2.21).

PROBLEM 4.2.1.   Assume that the hypotheses of Corollary 4.2.1 are satisfied. Set $\Gamma_2(t) = \alpha_3 + \tfrac{1}{2}\phi_2$ and $\Gamma_1(t) = \beta_3 + \alpha\alpha_3 + \tfrac{1}{2}\phi_1 + \tfrac{1}{2}\alpha\phi_2$. Let m and λ be constants satisfying $m \geqslant \exp\left[-\int_0^t \Gamma_2(s)\,ds\right]$ and $\lambda = m\|\Gamma_1\|_{L_1(R)}$. Let $u(t)$ be a solution of (4.2.36) for all real t and assume that there exist positive constants k and ε such that

$$\|u(t)\| = O[\exp(-kt)] \qquad \text{as} \quad t \to -\infty$$

and

$$\|u(t)\| = O[(\exp -[m^2 k + (m^2 + 1)\lambda + \varepsilon]t)] \qquad \text{as} \quad t \to +\infty.$$

Then, $u(t) \equiv 0$.

[Hint: Use Corollary 4.2.1 and Theorem 4.2.1 and argue as in Theorem 4.1.3.]

PROBLEM 4.2.2.   Assume that the hypotheses of Corollary 4.2.2 are satisfied and $\omega(t) \equiv 0$. Establish lower bounds for the solutions of (4.2.37) analogous to those of Theorem 4.1.2.

[Hint: Use Corollary 4.1.2 and Theorem 4.2.1.]

PROBLEM 4.2.3.   Assume that the hypotheses of Problem 4.2.2 are satisfied. Set

$$\gamma(t) = \max_{i=1,2,3} \gamma_i(t) \qquad \text{and} \qquad \beta = \max_{i=1,2,3} \beta_i(t).$$

We assume further that $T = \infty$ and

$$\gamma, t\beta, \phi_1 \in L_1[0, \infty) \qquad \text{and} \qquad t^{1/2}\gamma, t^{1/2}\phi_1 \in L_2[0, \infty).$$

Then the following convexity inequality holds for $0 \leqslant t_0 \leqslant t \leqslant t_1 < \infty$:

$$\log \|u(t)\| \leqslant \log \|u(t_0)\| \frac{\int_t^{t_1} \exp[\pm \int_{t_0}^s 4\gamma(r)\, dr]\, ds}{\int_{t_0}^{t_1} \exp[\pm \int_{t_0}^s 4\gamma(r)\, dr]\, ds}$$

$$+ \log \|u(t_1)\| \frac{\int_{t_0}^t \exp[\pm \int_{t_0}^s 4\gamma(r)\, dr]\, ds}{\int_{t_0}^{t_1} \exp[\pm \int_{t_0}^s 4\gamma(r)\, dr]\, ds} + K(t_0)$$

where one takes everywhere the negative sign if $\|u(t_0)\| \leqslant \|u(t_1)\|$ and the positive sign if $\|u(t_0)\| \geqslant \|u(t_1)\|$. Finally $K(t)$ is a nonnegative bounded function which depends only on γ, β, and ϕ_1 and $K(t) \to 0$ as $t \to \infty$.

[Hint: Use Corollary 4.2.2.]

4.3. Approximate Solutions, Bounds, and Uniqueness

Let us consider the evolution inequality

$$\|u' - A(t)u - f(t, u)\| \leqslant \phi(t, \|u\|), \qquad ' = d/dt \tag{4.3.1}$$

in the Banach space X, where $f \in C[R_+ \times X, X]$, $\phi \in [R_+ \times R_+, R_+]$, and for each $t \in R_+$ $A(t)$ is a linear operator with domain $D[A(t)] = D$ independent of t. We shall assume that for sufficiently small $h > 0$, the operators $R[h, A(t)] \equiv [I - hA(t)]^{-1}$ are well defined as bounded linear operators on X and that

$$\lim_{h \to 0_+} R[h, A(t)]x = x, \qquad x \in X. \tag{4.3.2}$$

The relation (4.3.2) is satisfied, if, for example, for each $t \in R_+$ $A(t)$ generates a strongly continuous semigroup. Notice that the operator $R[h, A(t)]$ is not exactly the resolvent of $A(t)$ but for sufficiently small $h > 0$, the number h^{-1} is in the resolvent set of $A(t)$.

DEFINITION 4.3.1.   A *solution* $u(t)$ of the evolution inequality (4.3.1) is a strongly continuously differentiable function $u: R_+ \to X$ with $u(t) \in D$ for $t \in R_+$ which satisfies (4.3.1) for all $t \in R_+ - S$ where S is a denumerable subset of R_+. If $\phi \equiv \varepsilon$, then $u(t)$ is said to be an *ε-approximate solution* of the evolution equation

$$u' = A(t)u + f(t, u). \tag{4.3.3}$$

THEOREM 4.3.1. Assume that

$$\| R[h, A(t)] x + hf(t, x)\| \leq \|x\| + h\psi(t, \|x\|) \qquad (4.3.4)$$

for all sufficiently small $h > 0$ and all $(t, x) \in R_+ \times X$, where $\psi \in C[R_+ \times R_+, R]$. Let $r(t, t_0, r_0)$ be the maximal solution of the scalar differential equation

$$r' = w(t, r) \qquad \text{and} \qquad r(t_0) = r_0 \geq 0 \qquad (4.3.5)$$

where $w(t, r) = \phi(t, r) + \psi(t, r)$ existing on $[t_0, \infty)$. Then any solution $u(t)$ of (4.3.1) satisfies the estimate

$$\|u(t)\| \leq r(t, t_0, r_0), \qquad t \geq t_0,$$

provided that $\|u(t_0)\| \leq r_0$.

Proof: Let $u(t)$ be any solution of (4.3.1). Define $m(t) = \|u(t)\|$. For $h > 0$ and sufficiently small we have, using (4.3.4)

$$m(t+h) = \|u(t+h)\| \leq \|u(t+h) - R[h, A(t)] u(t) - hf[t, u(t)]\|$$
$$+ m(t) + h\psi[t, m(t)]. \qquad (4.3.6)$$

Observe that

$$R[h, A(t)] u(t) = u(t) + hA(t) u(t) + h[R[h, A(t)] A(t) u(t) - A(t) u(t)].$$
$$(4.3.7)$$

In view of (4.3.1) and (4.3.2) the relations (4.3.6) and (4.3.7) lead to the scalar differential inequality

$$D_+ m(t) \leq w[t, m(t)] \qquad \text{and} \qquad m(t_0) \leq r_0$$

which implies the desired upper bound.

One can prove an analogous result for lower bounds.

THEOREM 4.3.2. Assume that

$$\| R[h, A(t)] x + hf(t, x)\|$$
$$\geq \|x\| - h\psi(t, \|x\|), \qquad h \text{ sufficiently small}, \quad (t, x) \in R_+ \times X, \quad (4.3.8)$$

where $\psi \in C[R_+ \times R_+, R]$. Let $\rho(t, t_0, \rho_0)$ be the minimal solution of the scalar differential equation

$$\rho' = -w(t, \rho) \qquad \text{and} \qquad \rho(t_0) = \rho_0 \geq 0$$

existing on $[t_0, \infty)$, where $w(t, \rho) = \phi(t, \rho) + \psi(t, \rho)$. Then any solution $u(t)$ of (4.3.1) satisfies the estimate

$$\|u(t)\| \geq \rho(t, t_0, \rho_0), \qquad t \geq t_0,$$

provided that $\|u(t_0)\| \geq \rho_0$.

Proof: Defining $m(t) = \|u(t)\|$ as before, it is easy to obtain the inequality

$$D_- m(t) \geq -w[t, m(t)] \qquad \text{and} \qquad m(t_0) \geq \rho_0$$

from which the stated estimate follows.

REMARK 4.3.1.   For various choices of ϕ and ψ, Theorems 4.3.1 and 4.3.2 extend many known results of ordinary differential equations in R^n to abstract differential equations.

(i)   Taking $\phi = \varepsilon$ and $\psi(t, r) = kr$, where k is a positive constant, Theorem 4.3.1 provides an upper estimate on the norm of ε-approximate solutions of (4.3.3), namely

$$\|u(t)\| \leq \|u(t_0)\| \exp[k(t-t_0)] + (\varepsilon/k)[\exp[k(t-t_0)] - 1], \qquad t \geq t_0$$

while Theorem 4.3.2 yields the lower estimate

$$\|u(t)\| \geq \|u(t_0)\| \exp[-k(t-t_0)] + (\varepsilon/k)[\exp[-k(t-t_0)] - 1], \qquad t \geq t_0.$$

(ii)   Suppose that $\phi \equiv 0$ and that $u(t)$ is a solution of (4.3.3) existing to the right of t_0. Let $\psi(t, r) = \lambda(t)r$ where $\lambda \in C[R_+, R]$ then from Theorems 4.3.1 and 4.3.2 the following upper and lower bounds follow:

$$\|u(t)\| \leq \|u(t_0)\| \exp \int_{t_0}^{t} \lambda(s)\, ds, \qquad t \geq t_0,$$

and

$$\|u(t)\| \geq \|u(t_0)\| \exp \int_{t_0}^{t} -\lambda(s)\, ds, \qquad t \geq t_0.$$

If, on the other hand, $\psi(t, r) = \lambda(t)g(r)$, where $g(r) > 0$ for $r > 0$, then

$$\|u(t)\| \leq G^{-1}\left[G(\|u(t_0)\|) + \int_{t_0}^{t} \lambda(s)\, ds \right]$$

and

$$\|u(t)\| \geq G^{-1}\left[G(\|u(t_0)\|) - \int_{t_0}^{t} \lambda(s)\, ds \right]$$

where $G(r) = \int_{r_0}^{r} ds/g(s)$ with $r_0 > 0$. These bounds hold as long as $G(\|u(t_0)\|) \pm \int_{t_0}^{t} \lambda(s)\, ds$ is in the domain of G^{-1}.

PROBLEM 4.3.1.   Assume that

   (i)   for each $t \in R_+$, $A(t)$ is the infinitesimal generator of a contraction semigroup in X;
   (ii)   for $(t, x) \in R_+ \times X$, $\|f(t, x)\| \leqslant g(t, \|x\|)$, where $g \in C[R_+ \times R_+, R_+]$;
   (iii) $r(t, t_0, r_0)$ is the maximal solution of the scalar differential equation

$$r' = g(t, r) \qquad \text{and} \qquad r(t_0) = r_0 \geqslant 0$$

existing on $[t_0, \infty)$.

If $u(t)$ is any solution of (4.3.3) existing on $[t_0, \infty)$ such that $\|u(t_0)\| \leqslant r_0$, then

$$\|u(t)\| \leqslant r(t, t_0, r_0), \qquad t \geqslant t_0.$$

Next we shall consider a general uniqueness result. For that purpose we shall assume that $f(t, 0) \equiv 0$ so that (4.3.1) has the identically zero solution. Here the functions ϕ, ψ and $w \equiv \phi + \psi$ are not required to be defined at $t = 0$. We then have the following.

THEOREM 4.3.3.   Let $B(t)$ be a positive continuous function on $0 < t < \infty$ with $B(0_+) = 0$. Assume that $w(t, 0) \equiv 0$ and that the only solution $r(t)$ of

$$r' = w(t, r) \tag{4.3.9}$$

existing on $0 < t < \infty$ and satisfying

$$\lim_{t \to 0_+} r(t)/B(t) = 0 \tag{4.3.10}$$

is the trivial solution. Then under the hypothesis (4.3.4) the only solution $u(t)$ of (4.3.1) satisfying the conditions $u(0) = 0$ and

$$\lim_{t \to 0} u(t)/B(t) = 0 \tag{4.3.11}$$

is the identically zero solution.

Proof: Let $u(t)$ be a solution of (4.3.1) satisfying (4.3.11). Define $m(t) = \|u(t)\|$ and note that $m(0) = 0$. To prove the theorem we have to show that $m(t) \equiv 0$ on R_+. Suppose, if possible, that $m(\sigma) > 0$ for some $\sigma > 0$. Let $r(t)$ be the minimal solution of (4.3.9) through $(\sigma, m(\sigma))$, existing on some interval to the left of σ. As far left of σ as $r(t)$ exists, it satisfies the inequality

$$r(t) \leqslant m(t). \tag{4.3.12}$$

To see that this is true, observe that

$$r' = w(t, r) + \varepsilon \qquad \text{and} \qquad r(\sigma) = m(\sigma) + \varepsilon \qquad (4.3.13)$$

has solutions $r(t, \varepsilon)$ for all sufficiently small $\varepsilon > 0$, existing as far to the left of σ as $r(t)$ exists and $\lim_{\varepsilon \to 0} r(t, \varepsilon) = r(t)$. Thus it is enough to prove that

$$r(t, \varepsilon) < m(t) \qquad (4.3.14)$$

for sufficiently small ε. If this inequality does not hold, let s be the least upper bound of numbers $t \leqslant \sigma$ for which (4.3.14) is false. Then it is easily seen that

$$m(s) = r(s, \varepsilon) \qquad \text{and} \qquad D_+ m(s) \geqslant r'(s, \varepsilon)$$

which contradict the inequality $D_+ m(t) \leqslant w[t, m(t)]$ obtained in the proof of Theorem 4.3.1. Hence (4.3.12) is valid. Next we prove that $r(t)$ can be continued up to $t = 0$. If $r(t_1) = 0$ for some t_1, such that, $0 < t_1 < \sigma$ the continuation can be effected by defining $r(t) \equiv 0$ for $0 < t < t_1$. Otherwise (4.3.12) ensures the possibility of continuation. Since $m(0) = 0$, we have $\lim_{t \to 0_+} r(t) = 0$ and we define $r(0) = 0$. Now we have a nontrivial solution of (4.3.9) on $0 \leqslant t \leqslant \sigma$ such that $r(\sigma) = m(\sigma)$ and $0 \leqslant r(t) \leqslant m(t)$. In view of (4.3.11) we have

$$0 \leqslant \lim_{t \to 0_+} r(t)/B(t)$$

$$\leqslant \lim_{t \to 0_+} m(t)/B(t) = 0$$

which by hypothesis implies that $r(t) \equiv 0$. This contradicts the fact that $r(\sigma) = m(\sigma) > 0$ and the proof is complete.

REMARK 4.3.2.   If $B(t) = t$, Theorem 4.3.3 is an extension of Kamke's uniqueness theorem which includes as special cases Nagumo's and Osgood's uniqueness conditions.

4.4. Application to Parabolic Equations

Let $(t, x) = (t, x_1, \ldots, x_n)$ be a generic point of $R \times R^n$. Set $\partial_i = \partial/\partial x_i$ and $\partial_x = (\partial_1, \ldots, \partial_n)$. Let Ω be a bounded domain of R^n with sufficiently smooth boundary such that

$$\int_\Omega u(\partial_i v)\, dx - \int_\Omega (\partial_i u) v\, dx = 0, \qquad u, v \in C_1[\overline{\Omega}, R], \quad u, v = 0 \quad \text{on } \partial\Omega.$$

($\partial\Omega$ is the boundary of Ω.) We shall denote by Γ the cylinder $[0, T) \times \Omega$ with $0 \leqslant T < +\infty$. On Γ we consider the parabolic operator $\partial_t - A$ where

$$A = \sum_{i,j=1}^{n} \partial_i a_{ij}(t,x) \partial_j + \sum_{i=1}^{n} b_i \partial_i + C$$

is a second-order elliptic operator with real coefficients such that

(i) for each $(\xi_1, \xi_2, ..., \xi_n) \in R^n$ with $k > 0$

$$\sum_{i,j=1}^{n} a_{ij} \xi_i \xi_j \geqslant k \sum_{i=1}^{n} \xi_1^2;$$

(ii) $a_{kj} = a_{jk} \in C_1[\overline{\Gamma}, R], \quad k,j = 1, ..., n,$

$$c, b_j \in C[\overline{\Gamma}, R], \quad j = 1, ..., n;$$

(iii) $$b \equiv \sum_{i=1}^{n} \partial_i b_i, \quad \partial_t b, \partial_t c \in C[\overline{\Gamma}, R];$$

(iv) there exists a real number λ_0 such that for all $(t,x) \in \Gamma$

$$c(t,x) - \tfrac{1}{2} b(t,x) \leqslant \lambda_0.$$

Under the above hypotheses we shall indicate how the previous results on abstract evolution inequalities apply to solutions of the inequality

$$\|\partial_t u - Au\|_{L_2(\Omega)} \leqslant \phi(t) \|u\|_{L_2(\Omega)} \tag{4.4.1}$$

where we assume that $u \in C_1[[0,T], L_2(\Omega)]$ for each $t \in [0,T)$, $u(t,x) \in C_2(\overline{\Omega})$, and $u(t,x) = 0$ on $\partial\Omega$.

First we make the change of variables $u = \exp(\lambda t) v$ with $\lambda \geqslant k + \lambda_0$. Then $v \in C_1[[0,T), L_2(\Omega)]$ for $v(t,x) = 0$ on $\partial\Omega$ and $v(t,x) \in C_2(\overline{\Omega})$ for each $t \in [0, T)$. In addition $v(t,x)$ satisfies the inequality

$$\|\partial_t v - (A - \lambda)v\|_{L_2(\Omega)} \leqslant \phi(t) \|v\|_{L_2(\Omega)}. \tag{4.4.2}$$

We shall prove that Assumptions 1–4 of Section 4.2 are satisfied for the operator $A - \lambda$.

Set

$$A_+ = A_0 + c - \tfrac{1}{2} b - \lambda, \qquad \delta A''_- = L + \tfrac{1}{2} b, \qquad A_-' = 0$$

where $A_0 = \sum_{i,j=1}^{n} \partial_i a_{ij} \partial_j$ and $L = \sum_{i=1}^{n} b_i \partial_i$. Here $\gamma_1 = \beta_1 = \alpha_1 = \alpha_2 = \alpha_3 = \phi_2 \equiv 0$ and therefore we need only verify Assumptions 2 and 3.

Let $w(t,x) \in D[A(t)]$ for each fixed t, $t \in [0, T)$, $w(t,x) \in C_2(\overline{\Omega})$, and $w(t,x) = 0$ on $\partial\Omega$ for each $t \in [0, T)$. Then $(A_+ w, w) = (A_0 w, w) + (C_0 w, w)$

where $C_0 = c - \frac{1}{2}b - \lambda \leqslant k$. Hence

$$|(A_+ w, w)| \geqslant k \left(\sum_{i=1}^{n} \|\partial_i w\|^2 + \|w\|^2 \right)$$

and

$$\|w\|_{H_1(\Omega)}^2 \equiv \sum_{i=1}^{n} \|\partial_i w\|^2 + \|w\|^2$$
$$\leqslant k^{-1} |(A_+ w, w)|$$
$$\leqslant k^{-1} \|A_+ w\| \, \|w\|.$$

Also

$$\|A''_- w\|^2 \leqslant 2\|Lw\|^2 + 2\|\tfrac{1}{2}bw\|^2$$
$$\leqslant C[\gamma(t)\|A_+ w\| \, \|w\| + \beta(t)\|w\|^2]$$
$$\equiv \delta\gamma_2 \|A_+ w\| \, \|w\| + \delta\beta_2 \|w\|^2$$

where $\gamma(t) = \sup_x \sum_{i=1}^{n} |b_i(t, x)|$, $\beta(t) = \sup_x |b(t, x)|^2$, $\delta\gamma_2(t) = C\gamma(t)$, and $\delta\beta_2(t) = \beta(t)$. Assumption 2 is therefore satisfied.

Finally we shall verify Assumption 3. For $u(t, x) \in C_1[[0, T), L_2(\Omega)]$ with $u(t, x) \in C_2(\overline{\Omega})$ and $u(t, x) = 0$ for $x \in \partial\Omega$ we have, for each $t \in [0, T)$

$$(d/dt)(A_+ u(t), u(t)) = -2 \sum_{i,j=1}^{n} (a_{ij} \partial_j u, \partial_i \partial_+ u) + 2(C_0 u, \partial_+ u)$$
$$- \sum_{i,j=1}^{n} ((\partial_t a_{ij}) \partial_i u, \partial_j u) + ((\partial_t C_0) u, u)$$
$$= 2(A_+ u, du/dt) - \sum_{i,j=1}^{n} ((\partial_t a_{ij}) \partial_i u, \partial_j u) + ((\partial_t C_0) u, u).$$

Hence

$$|(d/dt)(A_+ u(t), u(t)) - 2(A_+ u, du/dt)| \leqslant C[\Gamma(t)\|A_+ u\| \, \|u\| + B(t)\|u\|^2]$$
$$\equiv \delta\gamma_3 \|A_+ u\| \, \|u\| + \beta_3 \|u\|^2.$$

Now one can apply the results of Section 4.2 to obtain lower bounds for the solutions of (4.4.1).

PROBLEM 4.4.1. Let $T = \infty$. Assume that $\gamma_2, \beta_2, \gamma_3, \beta_3, \phi \in L_1[0, \infty) \cap L_\infty(0, \infty)$ then for some $\mu > 0$ and any solution of (4.4.1)

$$\|u(t, \cdot)\| \geqslant \|u(0, \cdot)\| \exp(-\mu t), \qquad t > 0.$$

[Hint: Use Problem 4.2.2.]

PROBLEM 4.4.2. Let u be a solution of the parabolic equation

$$\partial u/\partial t - \sum_{i,j=1}^{n} \partial_i [a_{ij}(x) \partial_j] u + a(x) u = 0 \qquad (4.4.3)$$

where a and the second partial derivatives of the a_{ij} are continuous and bounded in the closure of a domain Ω in R^n. Let u be a solution of (4.4.3) on Ω with zero Dirichlet boundary conditions such that for some positive constants C and β and for any real t

$$\|u(\cdot, t)\|_{L_2(\Omega)} \leqslant C \exp[\beta |t|].$$

Let U be a nonempty open subset of G. Assume that for any compact set K in U there exist positive constants C_1 and ε, depending on K, such that

$$\|u(\cdot, t)\|_{L_2(K)} \leqslant C_1 \exp[-(\beta + \varepsilon) t], \qquad t \geqslant 0.$$

Then $u \equiv 0$ on $U \times R$.

[Hint: Use Problem 4.2.1 with $m = 1$ and $\lambda = 0$.]

4.5. Notes

Theorems 4.1.1 and 4.1.2 are adapted from the work of Agmon and Nirenberg [2, 3]. See also Ogawa [54]. Theorem 4.1.3 is essentially due to Ogawa [55] and contains as a corollary Masuda's theorem [49]. Theorem 4.1.4 is a very special case of an analogous result in [3]. The results of Theorem 4.1.5 and Problem 4.1.2 are due to Agmon and Nirenberg [2]. Theorem 4.1.6 is adapted from Lakshmikantham [41]. All the results of Section 4.2 are taken from the work of Ladas and Lakshmikantham [36] which generalize and unify the corresponding results in [3] and [55]. For the contents of Section 4.3 see Lakshmikantham [41]. Further results can be found in Agmon and Nirenberg [3], Hurd [29], Ogawa [56] and Zaidman [80]. The material concerning the application given in Section 4.4 is taken from Agmon [1]. For earlier studies on this subject see also Agmon [1] and Lions [44].

Chapter 5

Nonlinear Differential Equations

5.0. Introduction

The contents of this chapter may perhaps be more interesting to those readers who have the flavor of differential equations in Euclidean spaces, since most of the information might, at first sight, appear not to apply to partial differential equations and unbounded operators. None the less it is here that the concept of monotonicity condition (Minty [51]) enters, out of which has grown a tremendous interest in the study of nonlinear semigroups and monotone operators. We devote Section 6.2 of the next chapter to the discussion of nonlinear semigroups where the importance of monotone operators unfolds itself. This chapter as a whole may be considered as developing the fundamental theory of nonlinear differential equations in Banach spaces.

We begin by giving some counterexamples to show that the classical Peano's existence theorem of differential equations in R^n, as well as the "continuation of solutions" theorem, cannot in general be extended to infinite-dimensional Banach spaces. We then offer a set of quite general sufficient conditions, more general than the monotonicity condition, that guarantee the existence and uniqueness of solutions of nonlinear abstract Cauchy problems, demonstrating the uniqueness part by an application to a parabolic partial differential equation. The purpose of Section 5.3 is to develop the nonlinear variation of constants formula, analogous to Alekseev's result [4]. As a preparation, a treatment of uniqueness of solutions and the continuity and differentiability with respect to initial conditions is given. Section 5.4 illustrates that the variation of constants formula is a convenient tool in studying the stability and asymptotic behavior of constantly acting perturbations. In Section 5.5 Chaplygin's method is exploited to approximate the solution of a nonlinear differential equation by a sequence of functions satisfying linear differential equations. After presenting a set of sufficient conditions for the global continuation of solutions of abstract Cauchy problems in Section 5.6 we extend the notion of asymptotic equilibrium to equations in a Banach space obtaining as an outcome an existence result for a terminal value Cauchy problem. Finally, using the extension of Lyapunov's second method, we consider in Section 5.7 several stability criteria of nonlinear evolution equations.

5.1. Counterexamples

Let X be a Banach space and $f(t, u)$ be a mapping from $[t_0, t_0 + a] \times X$ into X. Consider the initial value problem

$$du/dt = f(t, u), \qquad t_0 < t \leqslant t + a, \qquad (5.1.1)$$

$$u(t_0) = u_0, \qquad u_0 \in X. \qquad (5.1.2)$$

DEFINITION 5.1.1. A function $u: [t_0, t_0 + a] \to X$ is said to be a *solution* of (5.1.1) with initial value u_0 at $t = t_0$ if

(i) $\qquad\qquad\qquad u \in C[[t_0, t_0 + a], X];$

(ii) $\qquad\qquad\qquad u(t_0) = u_0;$

(iii) $u(t)$ is strongly differentiable in t for $t_0 < t \leqslant t_0 + a$ and satisfies (5.1.1) for $t_0 < t \leqslant t_0 + a$.

It is well known that in the case $X = R^n$, the n-dimensional Euclidean space, the continuity of f in a neighborhood of (t_0, u_0), alone, implies the existence of a local solution of (5.1.1) and (5.1.2). This is the classical Peano's theorem. This theorem cannot be generalized to the infinite-dimensional case. The following counterexample is known.

Consider the Banach space $X = (c_0)$ of real-valued sequences $u = \{\xi_n\}_{n=1}^{\infty}$ with $\lim_{n \to \infty} \xi_n = 0$ and norm $\|u\| = \sup_n |\xi_n|$. Define the function $f : X \to X$ by

$$f(u) = \{|\xi_n|^{\frac{1}{2}} + n^{-1}\}_{n=1}^{\infty}, \qquad u = \{\xi_n\}_{n=1}^{\infty} \in X.$$

The continuity of the real-valued function $\xi^{\frac{1}{2}}$ for $\xi \geq 0$ and the definition of the norm in X imply that the function $f(u)$ is continuous for all $u \in X$. However, the initial value problem

$$du/dt = f(u), \qquad u(0) = 0 \tag{5.1.3}$$

has no solution in X. In fact if $u(t) = \{\xi_n(t)\}_{n=1}^{\infty}$ were a solution of (5.1.3) the nth coordinate $\xi_n(t)$ should satisfy the scalar equation

$$\xi_n'(t) = |\xi_n(t)|^{\frac{1}{2}} + n^{-1} \tag{5.1.4}$$

and the initial condition

$$\xi_n(0) = 0. \tag{5.1.5}$$

From (5.1.4) $\xi_n(t)$ is strictly increasing in t and in view of (5.1.5) $\xi_n(t) > 0$ for $0 < t < \tau$ where τ is sufficiently small. Then from (5.1.4)

$$\xi_n'(t) > \xi_n^{\frac{1}{2}}(t), \qquad 0 < t < \tau$$

which leads to

$$\xi_n(t) \geq \tfrac{1}{4} t^2, \qquad 0 \leq t < \tau.$$

It is obvious now that no matter how small we choose τ the sequence $\{\xi_n(t)\}_{n=1}^{\infty}$ does not converge to zero as $n \to \infty$ which contradicts the hypothesis that $u(t)$ is a solution of (5.1.3) and in particular $u(t) \in X$. A similar argument holds to the left of $t = 0$. Thus, although the function f is continuous the initial value problem (5.1.3) has no solution in any open interval containing $t = 0$. At a first glance the preceding example seems to depend strongly on the properties of $X = (c_0)$ which is not reflexive (see [74]). However, another counterexample is known in a Hilbert space. (A Hilbert space is always reflexive.) We state it as follows.

PROBLEM 5.1.1.   Let H be the Hilbert space of real-valued sequences $u = \{\xi_n\}_{n=1}^{\infty}$ with $\|u\|^2 = \sum_{n=1}^{\infty} \xi_n{}^2$. Let P_n be the projections given by $P_n u = (0, \ldots, 0, \xi_{n+1}, \xi_{n+2}, \ldots)$ with $n = 1, 2, \ldots$ and $P_0 u = u$. For $(t, u) \in R \times H$ define

(i) $P(t) u = \begin{cases} 0, & t \leqslant 0 \\[2mm] u, & t \geqslant 1 \\[2mm] (2 - 2^n t) P_n u + (2^n t - 1) P_{n-1} u, \\[2mm] \qquad t \in [2^{-n}, 2^{-n+1}], \quad n = 1, 2, \ldots; \end{cases}$

(ii) $G(u) = u \|u\|^{-\frac{1}{2}}, \qquad u \neq 0 \quad \text{and} \quad G(0) = 0;$

(iii) $A(u) = (|\xi_1|, |\xi_2|, \ldots) \qquad \text{and} \qquad v = (2^{-1}, 2^{-2}, 2^{-3}, \ldots);$

(iv) $f(t, u) = G(P(t) A(u)) + P(t/2) v \max \{0, 4^{-1} t^2 - \|u\|\}.$

Then $f: R \times H \to H$ is continuous but the initial value problem

$$du/dt = f(t, u) \qquad \text{and} \qquad u(0) = 0 \tag{5.1.6}$$

has no solution in an open interval containing $t = 0$.

[Hint: The continuity of f follows from the continuity of P. Then assuming that (5.1.6) has a solution deduce a contradiction.]

REMARK 5.1.1.   The reason that the classical proof of Peano's theorem fails in an infinite-dimensional Banach space is that the closed unit ball in a Banach space is not necessarily compact. However, under the additional hypothesis that the function f is (locally) Lipschizian we can prove the (local) existence and uniqueness of solutions of (5.1.1) and (5.1.2).

THEOREM 5.1.1.   Define the rectangle

$$R_0 = \{(t, u) \in R \times X: |t - t_0| \leqslant \alpha, \quad \|u - u_0\| \leqslant \beta\}.$$

Let $f: R_0 \to X$ be continuous in t for each fixed u. Assume that $\|f(t, u)\| \leqslant M$ for $(t, u) \in R_0$ and $\|f(t, u_1) - f(t, u_2)\| \leqslant K \|u_1 - u_2\|$ for $(t, u_1), (t, u_2) \in R_0$ where K and M are nonnegative constants. Let α and β be positive constants satisfying $\alpha M \leqslant \beta$. Then there exists one and only one (strongly) continuously differentiable function $u(t)$ satisfying

$$du(t)/dt = f[t, u(t)], \qquad |t - t_0| \leqslant \alpha \tag{5.1.7}$$

and

$$u(t_0) = u_0. \tag{5.1.8}$$

Proof: It is clear that the continuity in t for each fixed u plus the Lipschitz condition implies that $f(t, u)$ is actually continuous in both variables. (In fact $\|f(t_1, u_1) - f(t_2, u_2)\| \leqslant \|f(t_1, u_1) - f(t_1, u_2)\| + \|f(t_1, u_2) - f(t_2, u_2)\| \leqslant K\|u_1 - u_2\| + \|f(t_1, u_2) - f(t_2, u_2)\|$.) We shall apply the classical method of successive approximations. Define

$$u_0(t) = u_0$$

and

$$u_n(t) = u_0 + \int_{t_0}^{t} f[s, u_{n-1}(s)]\, ds, \qquad |t - t_0| \leqslant \alpha,$$

the integral being taken relative to the strong topology. One can establish by induction that $u_n(t)$ is strongly continuous and that

$$\|u_n(t) - u_0\| \leqslant \beta, \qquad |t - t_0| \leqslant \alpha,$$

and also

$$\|u_n(t) - u_{n-1}(t)\| \leqslant MK^{n-1}(|t - t_0|^n/n!), \qquad n = 1, 2, \dots.$$

It follows that as $n \to \infty$ $u_n(t)$ converges uniformly in $|t - t_0| \leqslant \alpha$ to a strongly continuous function $u(t)$. Hence, as $n \to \infty$

$$\|f[t, u_n(t)] - f[t, u(t)]\| \leqslant K\|u_n(t) - u(t)\| \to 0,$$

uniformly in $|t - t_0| \leqslant \alpha$. Using Theorem 1.3.2 we obtain

$$u(t) = \lim_{n \to \infty} u_n(t) = u_0 + \lim_{n \to \infty} \int_{t_0}^{t} f[s, u_{n-1}(s)]\, ds = u_0 + \int_{t_0}^{t} f[s, u(s)]\, ds.$$

From this and Theorem 1.3.3 it is clear that $u(t)$ is continuously differentiable and satisfies (5.1.7) and (5.1.8). Finally we prove the uniqueness. Let $u(t)$ and $v(t)$ be two solutions of (5.1.7) and (5.1.8). Then

$$\|u(t) - v(t)\| = \left\| \int_{t_0}^{t} (f[s, u(s)] - f[s, v(s)])\, ds \right\| \leqslant K \int_{t_0}^{t} \|u(s) - v(s)\|\, ds$$

and by Gronwall's inequality

$$\|u(t) - v(t)\| \equiv 0.$$

The proof is therefore complete.

REMARK 5.1.2.   If X is a Banach algebra (see Appendix II) over a field F with unit e, the function $f(t, u) = au$ for $a, u \in X$ with fixed a satisfies the hypotheses of Theorem 5.1.1; therefore the Cauchy problem

$$du/dt = au \qquad \text{and} \qquad u(0) = e$$

has a unique solution. This solution is given by

$$u(t) = e + \sum_{n=1}^{\infty} (t^n a^n / n!)$$

and is taken as the definition of the *exponential function* $\exp(ta)$ in X.

Theorem 5.1.1 is a local result. In the case $X = R^n$ if $f \in C[[a, t_0 + a] \times X, X]$, not necessarily Lipschitzian, it is known that any solution $u(t)$ of

$$du/dt = f(t, u) \qquad \text{and} \qquad u(t_0) = u_0 \in X \qquad (5.1.9)$$

exists either on $[t_0, t_0 + a]$ or on $[t_0, \delta)$ with $0 < \delta \leqslant t_0 + a$ and $\|u(t)\| \to \infty$ as $t \to \delta$. If $f \in C[[t_0 - a, a] \times X, X]$, a similar result holds to the left of a. These results no longer hold if the underlying space X is an infinite dimensional Banach space.

We shall present a counterexample. Consider again the Banach space $X = (c_0)$ of real-valued sequences $u = \{\xi_n\}_{n=1}^{\infty}$ with $\lim_{n \to \infty} \xi_n = 0$ and norm $\|u\| = \sup_n |\xi_n|$. Let $e_n \in X$ be the vector $(0, ..., 0, 1, 0, ...)$ whose nth component is 1 and all others are equal to zero. We now define a sequence of functions $f_n \colon X \to X$ as follows: for $u = \{\xi_k\}_{k=1}^{\infty} \in X$, let

$$f_n(u) = [2\xi_n + 2\xi_{n+1} - 1]^+ (e_{n+1} - e_n)$$

where $[2\xi_n + 2\xi_{n+1} - 1]^+ = \max[0, 2\xi_n + 2\xi_{n+1} - 1]$.

Clearly, for each n, the function $f_n(u)$ is continuous and Lipschitzian in u. Also $f_n(u) \equiv 0$ for $\|u\| \leqslant \frac{1}{4}$ and $f_n(u) = e_{n+1} - e_n$ for $u = \lambda e_n + (1 - \lambda) e_{n+1}$ where λ is any real number. Let $\phi_n \in C[[(n+1)^{-1}, n^{-1}], R_+]$ be a sequence of functions such that

$$\phi((n+1)^{-1}) = \phi(n^{-1}) = 0 \qquad \text{and} \qquad \int_{(n+1)^{-1}}^{n^{-1}} \phi_n(t) \, dt = 1, \qquad n = 1, 2, \dots.$$

Define the function $f \colon (-\infty, 1) \times X \to X$ by

$$f(t, u) = \begin{cases} 0, & t \leqslant 0, & u \in X, \\ \phi_n(t) f_n(u), & t \leqslant [(n+1)^{-1}, n^{-1}], & u \in X. \end{cases}$$

One can now prove that the function $f(t, u)$ is locally Lipschitzian in u and continuous for all points (t, u) with $t \neq 0$. f is also continuous at any point $(0, a)$ where $a = \{\alpha_n\}_{n=1}^{\infty} \in X$. In fact since $\lim_{n \to \infty} \alpha_n = 0$, there exists an index N such that $|\alpha_n| \leqslant \frac{1}{8}$ for all $n \geqslant N$. Consider the ball $B = \{u \in X:$

$\|u-a\| \leqslant \frac{1}{8}$. If $u = \{\xi_n\}_{n=1}^{\infty} \in B$, then $|\xi_n| \leqslant \frac{1}{4}$ for all $n \geqslant N$ and so $f_n(u) = 0$ for all $n \geqslant N$. Therefore, $f(t,u) = 0$ for $(t,u) \in [0, N^{-1}] \times B$ which establishes our assertion. Next define the sequence of functions $v_n: [0,1] \to X$ for $n = 1, 2, \ldots$ by

$$v_1(t) = \begin{cases} e_1 + (e_2 - e_1) \displaystyle\int_t^1 \phi_1(s)\, ds, & \frac{1}{2} \leqslant t \leqslant 1 \\ 0, & t < \frac{1}{2} \end{cases}$$

$$v(t) = \begin{cases} e_n + (e_{n+1} - e_n) \displaystyle\int_t^{n-1} \phi_n(s)\, ds, & (n+1)^{-1} \leqslant t < n^{-1} \\ 0, & t \in [0,1] - [(n+1)^{-1}, n^{-1}) \end{cases}$$

and also the function $u: (0,1] \to X$ by

$$u(t) = \sum_{n=1}^{\infty} v_n(t). \qquad (5.1.10)$$

The series (5.1.10) always makes sense for $t \in (0,1]$ because it has only a finite number of nonzero terms. For $t \in [(n+1)^{-1}, n^{-1})$ we have

$$u(t) = e_n + (e_{n+1} - e_n) \int_t^{n-1} \phi_n(s)\, ds. \qquad (5.1.11)$$

From (5.1.11) it results immediately that $\|u(t)\| \leqslant 1$ for $0 < t \leqslant 1$, that $u(t)$ is differentiable for $0 < t \leqslant 1$, and that

$$u'(t) = -(e_{n+1} - e_n)\phi_n(t)$$
$$= -f_n[u(t)]\phi_n(t)$$
$$= -f[t, u(t)].$$

Clearly $u(1) = e_1$.

Hence, we have established that $u(t)$ satisfies the initial value problem

$$du/dt = -f(t, u), \qquad 0 < t \leqslant 1 \qquad (5.1.12)$$

and

$$u(1) = e_1 \qquad (5.1.13)$$

with $\|u(t)\| \leqslant 1$ for $0 < t \leqslant 1$. However, $u(t)$ does not tend to a limit as $t \to 0$. In fact $u(n^{-1}) = e_n$ and e_n does not have a limit as $n \to \infty$.

5.2. Existence and Uniqueness

In Section 5.1 we have seen that the mere continuity of f is not enough to guarantee the existence of a solution of the initial value problem (5.1.1) and (5.1.2) in an infinite-dimensional Banach space X. On the other hand, if f is (locally) Lipschitzian in X, the existence and uniqueness of (local) solutions is assured and the proof is identical with the proof given in R^n. First we shall present a set of quite general conditions on f which guarantee that the Cauchy problem

$$du/dt = f(t, u), \qquad t_0 \leqslant t \leqslant t_0 + a \qquad (5.2.1)$$

and

$$u(t_0) = u_0 \qquad (5.2.2)$$

is *well posed*, that is, solutions of (5.2.1) and (5.2.2) exist, are unique, and depend continuously on their initial data.

THEOREM 5.2.1.   Assume that

(i) $f \in C[[t_0, t_0 + a] \times S_b, X]; \qquad S_b = \{u \in X : \|u - u_0\| \leqslant b\}$

and

$$\|f(t, u)\| \leqslant M, \qquad (t, u) \in [t_0, t_0 + a] \times S_b;$$

(ii)   there exists a functional $V \in C[[t_0, t_0 + a] \times S_b \times S_b, R_+]$ such that

(a) $\qquad\qquad\qquad V(t, u, v) > 0, \qquad u \neq v;$

(b) $\qquad\qquad\qquad V(t, u, v) \equiv 0, \qquad u = v;$

(c)   if $\lim_{n \to \infty} V(t, u_n, v_n) = 0$ for each $t \in [t_0, t_0 + a]$, then $\lim_{n \to \infty} (u_n - v_n)$ $= 0;$

(d) $V(t, u, v)$ is continuously Fréchet differentiable and

$$\partial V(t, u, v)/\partial t + [\partial V(t, \cdot, v)/\partial u] f(t, u) + [\partial V(t, u, \cdot)/\partial v] f(t, v) \leqslant 0;$$

(e)   for any positive number K the functions $\partial V(t, u, v)/\partial t$, $[\partial V(t, \cdot v)/\partial u] x$, and $[\partial V(t, u, \cdot)/\partial v] x$ are continuous in (u, v) uniformly for $(t, u, v) \in [t_0, t_0 + a] \times S_b \times S_b$ and $x \in X$ with $\|x\| \leqslant K$.

Then (5.2.1) and (5.2.2) possesses a unique solution on $[t_0, t_0 + \alpha]$ where $\alpha > 0$ satisfies $\alpha M \leqslant b$. Moreover, the solution depends continuously on (t_0, u_0).

Proof: Let $\Delta: t_0 < t_1 < \cdots < t_n = t_0 + \alpha$ be a subdivision of $[t_0, t_0 + \alpha]$. We define the function $\phi_\Delta: [t_0, t_0 + \alpha] \to X$ as follows:

$$\phi_\Delta(t_0) = u_0,$$

$$\phi_\Delta(t) = \phi_\Delta(t_{k-1}) + \int_{t_{k-1}}^{t} f[s, \phi_\Delta(t_{k-1})] \, ds,$$

$$t_{k-1} < t \leqslant t_k, \quad k = 1, 2, \ldots, n.$$

It is clear from the definition that $\phi_\Delta(t)$ is well defined on $[t_0, t_0 + \alpha]$ and

$$\phi_\Delta(t_{k-1}) = \phi_\Delta(t_{k-2}) + \int_{t_{k-2}}^{t_{k-1}} f[s, \phi_\Delta(t_{k-2})] \, ds, \qquad k = 2, 3, \ldots, n.$$

Hence

$$\phi_\Delta(t) = \phi_\Delta(t_0) + \int_{t_0}^{t_1} f[s, \phi_\Delta(t_0)] \, ds + \cdots + \int_{t_{k-2}}^{t_{k-1}} f[s, \phi_\Delta(t_{k-2})] \, ds$$

$$+ \int_{t_{k-1}}^{t} f[s, \phi_\Delta(t_{k-1})] \, ds$$

$$\equiv u_0 + \int_{t_0}^{t} f_\Delta(s) \, ds \tag{5.2.3}$$

where the function $f_\Delta: [t_0, t_0 + \alpha] \to X$ and

$$f_\Delta(s) = f[s, \phi(t_k)], \qquad t_k < s \leqslant t_{k+1}.$$

Notice that

$$\phi_\Delta'(t) = f(t, \phi_\Delta(t_{k-1})), \qquad t_{k-1} < t < t_k. \tag{5.2.4}$$

In view of (5.2.3) and the definition of α

$$\|\phi_\Delta(t) - u_0\| \leqslant M(t - t_0) \leqslant M\alpha \leqslant b, \qquad t_0 \leqslant t \leqslant t_0 + \alpha$$

and therefore $\phi_\Delta(t) \in S_b$ for $t_0 \leqslant t \leqslant t_0 + \alpha$. Let $\Delta, \tilde{\Delta}$ be two subdivisions of $[t_0, t_0 + \alpha]$ and $\phi_\Delta(t), \phi_{\tilde{\Delta}}(t)$ the corresponding functions. If t is not a subdivision point of either Δ or $\tilde{\Delta}$, $t_{k-1} < t \leqslant t_k$ and $\tilde{t}_{j-1} < t \leqslant \tilde{t}_j$, then (using Lemma 1.6.4)

$$(d/dt)(V[t, \phi_\Delta(t), \phi_{\tilde{\Delta}}(t)])$$

$$= \partial V/\partial t + (\partial V/\partial u) \phi_\Delta'(t) + (\partial V/\partial v) \phi_{\tilde{\Delta}}'(t)$$

$$= \partial V/\partial t + (\partial V/\partial u) f[t, \phi_\Delta(t_{k-1})] + (\partial V/\partial v) f[t, \phi_{\tilde{\Delta}}(\tilde{t}_{j-1})]$$

$$= [\partial V/\partial t - \partial \tilde{V}/\partial t] + \partial \tilde{V}/\partial t$$
$$+ [(\partial V/\partial u - \partial \tilde{V}/\partial u) + \partial \tilde{V}/\partial u] f(t, \phi_\Delta(t_{k-1}))$$
$$+ [(\partial V/\partial v - \partial \tilde{V}/\partial v) + \partial \tilde{V}/\partial v] f(t, \phi_{\tilde{\Delta}}(\tilde{t}_{j-1})) \qquad (5.2.5)$$

where

$$\partial V/\partial t = (\partial V/\partial t)[t, \phi_\Delta(t), \phi_{\tilde{\Delta}}(t)], \quad \partial \tilde{V}/\partial t = (\partial V/\partial t)[t, \phi_\Delta(t_{k-1}), \phi_{\tilde{\Delta}}(\tilde{t}_{j-1})];$$

$$\partial V/\partial u = (\partial V/\partial u)[t, \cdot, \phi_{\tilde{\Delta}}(t)], \quad \partial \tilde{V}/\partial u = (\partial V/\partial u)[t, \cdot, \phi_{\tilde{\Delta}}(\tilde{t}_{j-1})];$$

$$\partial V/\partial v = (\partial V/\partial v)[t, \phi_\Delta(t), \cdot], \quad \partial \tilde{V}/\partial v = (\partial V/\partial v)[t, \phi_\Delta(t_{k-1}), \cdot].$$

In view of Hypothesis (ii)(e), for any $\varepsilon > 0$ there exists a $\delta > 0$ such that if we take $|\Delta| \equiv \max_k (t_k - t_{k-1}) < \delta$ and $|\tilde{\Delta}| = \max_j (\tilde{t}_j - \tilde{t}_{j-1}) < \delta$ then we have

$$\partial V/\partial t - \partial \tilde{V}/\partial t < \varepsilon/3 \qquad (5.2.6)$$

$$(\partial V/\partial u - \partial \tilde{V}/\partial u) f[t, \phi_\Delta(t_{k-1})] < \varepsilon/3 \qquad (5.2.7)$$

$$(\partial V/\partial v - \partial \tilde{V}/\partial v) f[t, \Phi_{\tilde{\Delta}}(\tilde{t}_{j-1})] < \varepsilon/3 \qquad (5.2.8)$$

and from Hypothesis (ii)(d)

$$\partial \tilde{V}/\partial t + (\partial \tilde{V}/\partial u) f[t, \phi_\Delta(t_{k-1})] + (\partial \tilde{V}/\partial v) f[t, \phi_{\tilde{\Delta}}(\tilde{t}_{j-1})] \leq 0. \qquad (5.2.9)$$

From (5.2.5) and the inequalities (5.2.6)–(5.2.9) it follows that

$$(d/dt)(V[t, \phi_\Delta(t), \phi_{\tilde{\Delta}}(t)]) < \varepsilon. \qquad (5.2.10)$$

Integrating (5.2.10) from t_0 to t and using Hypothesis (ii)(b), we obtain

$$V[t, \phi_\Delta(t), \phi_{\tilde{\Delta}}(t)] \leq \varepsilon(t - t_0)$$
$$\leq \varepsilon \alpha. \qquad (5.2.11)$$

The estimate (5.2.11) together with Hypotheses (ii)(a) and (ii)(b), and the completeness of X implies that there exists a function $u(t)$ such that

$$\lim_{|\Delta| \to 0} \phi_\Delta(t) = u(t), \qquad t \in [t_0, t_0 + \alpha].$$

Clearly $u(t) \in S_b$. Now, fix $t \in [t_0, t_0 + \alpha]$. Since for each subdivision Δ there exists a k such that $t_{k-1} < t \leq t_k$ and

$$|\phi_\Delta(t_{k-1}) - \phi_\Delta(t)| \leq M(t - t_{k-1})$$
$$\leq M|\Delta|.$$

It follows that $\lim_{|\Delta| \to 0} \phi_\Delta(t_{k-1}) = u(t)$. Then $f[t, \phi_\Delta(t_{k-1})] = f_\Delta(t) \to f[t, u(t)]$ as $|\Delta| \to 0$. By the dominated convergence theorem and (5.2.3) we obtain

$$u(t) = u_0 + \int_{t_0}^t f[s, u(s)] \, ds.$$

The existence part of Theorem 5.2.1 is proved.

To prove uniqueness, assume that $u(t)$ and $v(t)$ are two solutions of (5.2.1) and (5.2.2). Then from Hypothesis (ii)(d)

$$(d/dt)V[t, u(t), v(t)] = \partial V/\partial t + (\partial V/\partial u) u'(t) + (\partial V/\partial v) v'(t)$$
$$= \partial V/\partial t + (\partial V/\partial u) f[t, u(t)] + (\partial V/\partial v) f[t, v(t)] \leq 0.$$

Integrating from t_0 to t we get

$$V[t, u(t), v(t)] - V[t_0, u_0, u_0] \leq 0$$

which on the strength of Hypotheses (ii)(b) and (ii)(a) yields $u(t) \equiv v(t)$. Finally, we shall prove the continuous dependence of the solutions with respect to initial conditions. Let $u_1(t) = u(t, t_1, u_1)$ and $u_2(t) = u(t, t_2, u_2)$ be the solutions of (5.2.1) through (t_1, u_1) and (t_2, u_2), respectively. Then in view of Hypothesis (ii)(d)

$$(d/dt)V[t, u_1(t), u_2(t)] \leq 0$$

and integrating from t_1 to t

$$0 \leq V[t, u_1(t), u_2(t)] \leq V[t_1, u_1, u_2(t_1)]. \tag{5.2.12}$$

Let $(t_1, u_1) \to (t_2, u_2)$. Since V and $u_2(t)$ are continuous we get

$$V[t_1, u_1, u_2(t_1)] \to V[t_2, u_2, u_2(t_2)] = 0.$$

Taking limits on both sides of (5.2.12) as $(t_1, u_1) \to (t_2, u_2)$ and using Hypothesis (ii)(c) the desired result follows. The proof is complete.

An interesting special case of Theorem 5.2.1 in a Hilbert space is the following:

COROLLARY 5.2.1. Let X be a Hilbert space and let Hypothesis (i) of Theorem 5.2.1 be satisfied. Furthermore, assume that $-f$ is a *monotonic function*, that is, there exists a constant M such that

$$\text{Re}[f(t, u) - f(t, v), u - v] \leq M \|u - v\|^2, \qquad t_0 \leq t \leq t_0 + \alpha, \quad u, v \in X. \tag{5.2.13}$$

Then the conclusion of Theorem 5.2.1 is satisfied.

Proof: It suffices to exhibit a functional $V(t, u, v)$ satisfying Hypothesis (ii) of Theorem 5.2.1. In fact, set

$$V(t, u, v) = \exp(-2Mt)\|u - v\|^2.$$

Then from the results of Section 1.6 and the monotonicity of $-f$ we obtain

$$\partial V(t, u, v)/\partial t + [\partial V(t, \cdot, v)/\partial u]f(t, u) + [\partial V(t, u, \cdot)/\partial v]f(t, v)$$

$$= -M\exp(-2Mt)\|u - v\|^2 + 2\exp(-2Mt)\operatorname{Re}(f(t, u), u - v)$$

$$- 2\exp(-2Mt)\operatorname{Re}(f(t, v), u - v)$$

$$= 2\exp(-2Mt)[\operatorname{Re}(f(t, u) - f(t, v), u - v) - M\|u - v\|^2] \leqslant 0$$

and Hypothesis (ii)(d) is satisfied. Clearly all other conditions in (ii) are satisfied and the proof is complete.

REMARK 5.2.1. Let $X = R^1$. The function

$$f(t, u) = \begin{cases} 1 - \sqrt{u}, & u \geqslant 0 \\ 1, & u < 0 \end{cases}$$

does not satisfy a Lipschitz condition but does satisfy the monotonicity condition (5.2.13) with $M = 0$. On the other hand, the function

$$f(t, u) = \begin{cases} 1 + \sqrt{u}, & u \geqslant 0 \\ 1, & u < 0 \end{cases}$$

does not satisfy the monotonicity condition (5.2.13) but there does exist a functional $V(t, u, v)$ satisfying all the conditions of Theorem 5.2.1. Indeed, take

$$V(t, u, v) = \begin{cases} [\sqrt{u} - \sqrt{v} - \log(1 + \sqrt{u}) + \log(1 + \sqrt{v})]^2, & u \geqslant 0, \quad v \geqslant 0 \\ [\sqrt{u} - \log(1 + \sqrt{u}) - \tfrac{1}{2}v]^2, & u \geqslant 0, \quad v < 0 \\ [\tfrac{1}{2}u - \sqrt{v} + \log(1 + \sqrt{v})]^2, & u < 0, \quad v \geqslant 0 \\ \tfrac{1}{4}(u - v)^2, & u < 0, \quad v < 0. \end{cases}$$

Next we shall prove a general uniqueness theorem in a normed space X. Let $I = [t_0, t_0 + a]$ and for each $t \in I$, let $D(t)$ be a subset of X. Define

$$D = \{(t, u): t \in I \quad \text{and} \quad u \in D(t)\}.$$

Consider the initial value problem

$$du/dt = f(t, u) \qquad (5.2.14)$$

$$u(t_0) = u_0 \in D(t_0) \qquad (5.2.15)$$

where $f: D \to X$. (f is not necessarily continuous.)

THEOREM 5.2.2.   Assume that there exists a functional $V \in C[I \times D \times D, X]$ satisfying Hypotheses (ii)(a), (ii)(b), and (ii)(d) of Theorem 5.2.1. Then (5.2.14) and (5.2.15) has at most one solution. Furthermore, if hypothesis (ii)(c) is satisfied, the solution depends continuously on the initial conditions.

The proof is identical with the uniqueness and the continuous dependence proof of Theorem 5.2.1 and we shall omit it.

Finally we apply Theorem 5.2.2 to prove a known uniqueness result for the solutions of a parabolic partial differential equation.

EXAMPLE 5.2.1.   Consider the parabolic equation

$$\partial u/\partial t = \partial^2 u/\partial x^2 + F(t, x, u) \qquad (5.2.16)$$

on a region bounded by $t = t_0$, $t = t_0 + a$, $x = \lambda_1(t)$, and $x = \lambda_2(t)$ where $\lambda_1(t)$ and $\lambda_2(t)$ are differentiable on $I = [t_0, t_0 + a]$ and $\lambda_1(t) < \lambda_2(t)$ for $t \in I$. The initial and boundary conditions are $u = g(x)$ on $t = t_0$, $u = h_1(t)$ on $x = \lambda_1(t)$, and $u = h_2(t)$ on $x = \lambda_2(t)$ where g, h_1 and h_2 are continuous, and $g[\lambda_1(t_0)] = h_1(t_0)$ and $g[\lambda_2(t_0)] = h_2(t_0)$. Assume that for some constant K

$$F(t, x, u_1) - F(t, x, u_2) \leqslant K(u_1 - u_2), \qquad u_1 > u_2.$$

Then (5.2.16) has a unique solution. To prove this let $X = L_2(R)$. For each $t \in I$ define $D(t)$ to be the space of functions $u = u(x) \in X$ which are continuous on $[\lambda_1(t), \lambda_2(t)]$, belong to C_2 on $(\lambda_1(t), \lambda_2(t))$, vanish outside $[\lambda_1(t), \lambda_2(t)]$, and take the values $h_1(t)$ and $h_2(t)$ at $x = \lambda_1(t)$ and $x = \lambda_2(t)$, respectively. For $(t, u) \in I \times D(t)$ define

$$f(t, u) = d^2 u(x)/dx^2 + F(t, x, u).$$

Then, (5.2.16) together with the initial and boundary conditions is equivalent to the initial value problem

$$du/dt = f(t, u) \qquad (5.2.17)$$

$$u(t_0) = g \qquad (5.2.18)$$

Define

$$V(t, u, v) = \exp(-2Kt) \int_{\lambda_1(t)}^{\lambda_2(t)} |u(x) - v(x)|^2 \, dx. \tag{5.2.19}$$

We shall now verify that the conditions of Theorem 5.2.2 are satisfied for the system (5.2.17) and (5.2.18) and the choice (5.2.19) of the functional V. In fact the conditions (ii)(a) and (ii)(c) are obvious. Let us verify Hypothesis (ii)(d). From (5.2.19) we have

$$\partial V / \partial t + (\partial V / \partial u) f(t, u) + (\partial V / \partial v) f(t, v)$$

$$= -2K \exp(-2Kt) \int_{\lambda_1(t)}^{\lambda_2(t)} |u(x) - v(x)|^2 \, dx$$

$$+ \exp(-2Kt)(|u[\lambda_2(t)] - v[\lambda_2(t)]|^2 \lambda_2'(t) - |u[\lambda_1(t)] - v[\lambda_1(t)]|^2$$

$$\times \lambda_1'(t))$$

$$+ \exp(-2Kt) \int_{\lambda_1(t)}^{\lambda_2(t)} 2[u(x) - v(x)] f(t, u) \, dx$$

$$- \exp(-2Kt) \int_{\lambda_1(t)}^{\lambda_2(t)} 2[u(x) - v(x) f(t, v) \, dx$$

$$\equiv J_1 + J_2 + J_3 + J_4.$$

Since $u[\lambda_i(t)] = v[\lambda_i(t)] = h_i(t)$ for $i = 1, 2$ it follows that $J_2 = 0$. From the definition of $f(t, u)$ and an integration by parts we obtain

$$J_3 + J_4 = 2 \exp(-2Kt) \int_{\lambda_1(t)}^{\lambda_2(t)} [u(x) - v(x)][u_{xx} + f(t, x, u) - v_{xx} - f(t, x, v)] \, dx$$

$$= 2 \exp(-2Kt) \int_{\lambda_1(t)}^{\lambda_2(t)} [u(x) - v(x)][f(t, u) - f(t, v)] \, dx$$

$$- 2 \exp(-2Kt) \int_{\lambda_1(t)}^{\lambda_2(t)} |u_x - v_x|^2 \, dx$$

$$\leqslant 2 \exp(-2Kt) \int_{\lambda_1(t)}^{\lambda_2(t)} [u(x) - v(x)][f(t, u) - f(t, v)] \, dx.$$

From these observations and the hypothesis on F we conclude that Hypothesis (ii)(d) is valid and therefore (5.2.16) has at most one solution.

PROBLEM 5.2.1 (Nagumo-type uniqueness). Let H be a Hilbert space and $f: R_+ \times H \to H$ satisfy the condition

$$\text{Re}(f(t, u) - f(t, v), u - v) \leqslant (2t)^{-1} \|u - v\|^2, \qquad t > 0, \quad u, v \in H.$$

Then the Cauchy problem

$$du/dt = f(t, u) \qquad \text{and} \qquad u(0) = u_0 \qquad (5.2.20)$$

has at most one solution.

[Hint: If $u(t)$ and $v(t)$ are two solutions define $m(t) = \|u(t) - v(t)\|^2$ and prove that $m'(t) \leq (2t)^{-1} m(t)$ with $m(0) = 0$ and $m'(0) = 0$.]

PROBLEM 5.2.2   (Osgood-type uniqueness).   Let $w: [0, \infty) \to R_+$ be strictly increasing and such that $w(0) = 0$ and for each T such that $0 < T < \infty$

$$\lim_{|\varepsilon| \to 0} \int_\varepsilon^T ds/w(s) = \infty.$$

Assume that $f: R_+ \times H \to H$ and for all $t > 0$ and $u, v \in H$

$$2 \operatorname{Re}(f(t, u - f(t, v), u - v) \leq w(\|u - v\|^2).$$

Then (5.2.20) has at most one solution.

[Hint: Set $m(t) = \|u(t) - v(t)\|^2$; then $m'(t) \leq w[m(t)]$.]

PROBLEM 5.2.3.   Let X be a reflexive Banach space. Then the conclusion of Theorem 5.2.1 remains valid if we replace the continuity of f by *demicontinuity* (that is, f is continuous from $[t_0, t_0 + a] \times X$ with the strong topology, into X with the weak topology) and differentiation is understood in the weak sense.

[Hint: Establish that $f_\Delta(t)$ converges weakly to $f[t, \phi(t)]$ as $|\Delta| \to 0$.]

PROBLEM 5.2.4.   Consider the nonlinear evolution equation

$$du/dt = A(t)u + f(t, u), \qquad t_0 < t \leq t_0 + a \qquad (5.2.21)$$

where the operators $A(t)$ satisfy Hypotheses 1–3 of Section 3.1. Assume that the hypotheses (i) and (ii) of Theorem 5.2.1 are satisfied. Furthermore, assume that $f(t, u)$ is Hölder continuous in t and u and that for $u, v \in D$ and $t \in [t_0, t_0 + a]$

$$[\partial V(t, \cdot, v)/\partial u] A(t) u + [\partial V(t, u, \cdot)/\partial v] A(t) v \leq 0.$$

Then, the evolution equation (5.2.21) has a mild solution, on $[t_0, t_0 + \alpha]$ where $0 < \alpha \leq a$, through the point (t_0, u_0) with $u_0 \in D$.

[Hint: Define as in Theorem 5.2.1 $\phi_\Delta(t_0) = u_0$ and

$$\phi_\Delta(t) = U(t, t_{k-1})\phi_\Delta(t_{k-1}) + \int_{t_{k-1}}^{t} U(t, s)f[s, U(s, t_{k-1})]\phi_\Delta(t_{k-1})\,ds$$

for $t_{k-1} \leqslant t \leqslant t_k$ where U is the fundamental solution constructed in Section 3.4.]

5.3. Nonlinear Variation of Constants Formula

In this section, we consider the nonlinear abstract Cauchy problem

$$du/dt = f(t, u) \quad \text{and} \quad u(t_0) = u_0 \tag{5.3.1}$$

where $f: R_+ \times X \to X$ is a given function and X a Banach space. Our aim is to develop the variation of constants formula with respect to (5.3.1) and its perturbation

$$dv/dt = f(t, v) + F(t, v) \quad \text{and} \quad v(t_0) = v_0 \tag{5.3.2}$$

where $F: R_+ \times X \to X$. As we shall see later, this result is a convenient tool in discussing the properties of solutions of the perturbed system (5.3.2) including the preservation of stability properties under constantly acting perturbations. First, it is necessary to study the uniqueness of solutions, their continuity and differentiability with respect to initial conditions (t_0, u_0), and to show that the Fréchet derivatives of the solutions $u(t, t_0, u_0)$ of (5.3.1) with respect to initial values exist and satisfy the equation of variation of (5.3.1) along the solution $u(t, t_0, u_0)$. Our treatment rests on the existence of an admissible functional in X, a mild one-sided estimate of f, and the theory of scalar differential inequalities. By $S(x_0, r)$ we shall denote the sphere $\{x \in X: \|x - x_0\| \leqslant r\}$. For a function $f: R_+ \times X \to X$ the Fréchet derivative with respect to x, if it exists, is denoted by $f_x(t, \cdot)$ and as we have seen in Section 1.6 it belongs to $B(X)$. The notation $w(h) = O(\|h\|)$ for $h \in X$ stands for a vector in X satisfying the condition

$$\lim_{\|h\| \to 0} w(h)/\|h\| = 0.$$

DEFINITION 5.3.1. A (nonlinear) continuous functional $\Phi: X \to R_+$ is said to be *admissible* in X if the following conditions are satisfied:

(i) $\Phi(x) > 0, \quad x \in X, \quad x \neq 0 \quad \text{and} \quad \Phi(0) = 0;$

(ii)   if $\lim_{n \to \infty} x_n = 0$ for $x_n \in X$, then $\lim_{n \to \infty} x_n = 0;$

(iii)   there exists a mapping $M: X \times X \to R$ such that $M[x,h]$ is continuous in h, uniformly with respect to x in any sphere $S(x_0, r)$, and satisfying the properties

(a) $\Phi(x+h) - \Phi(x) \leqslant M[x,h] + 0(\|h\|), \qquad x, h \in X;$

(b) $M[x, \lambda h] = \lambda M[x,h], \qquad \lambda \geqslant 0, \qquad x, h \in X;$

(c) $M[x, h_1 + h_2] \leqslant M[x, h_1] + M[x, h_2], \qquad x, h_1, h_2 \in X.$

EXAMPLE 5.3.1.   (i)   The functional $\Phi(x) = \|x\|$ is admissible in any Banach space X with $M[x,h] = \|h\|$.

(ii)   The function $\Phi(x) = \sum_{i=1}^{n} |x_i|$ is admissible in R^n where for $x = (x_1, \ldots, x_n)$, $h = (h_1, h_2, \ldots, h_n) \in R^n$.

$$M[x,h] = \sum_{i=1}^{n} M[x_i, h_i], \qquad M[x_i, h_i] = \begin{cases} h_i \operatorname{sgn} x_i, & x_i \neq 0 \\ |h_i|, & x_i = 0. \end{cases}$$

(iii)   The functional $\Phi(x) = (x, x)$ is admissible in any Hilbert space H with $M[x,h] = 2\operatorname{Re}(x,h)$.

The following lemma shows that the functional $M[x, \cdot]$ is bounded.

LEMMA 5.3.1.   Let Φ be an admissible functional in X. Then for any sphere $S = S(x_0, r)$ there is a constant $K(r)$ such that

$$|M[x,h]| < K(r)\|h\|, \qquad x \in S, \quad h \in X.$$

Proof: In view of the continuity of $M[x,h]$ in h, uniformly with respect to $x \in S$, and the fact that $M[x,0] = 0$, it follows that, given $\varepsilon > 0$ there exists a $\delta = \delta(\varepsilon, r)$ such that $\|\tilde{h}\| \leqslant \delta$ implies $|M[x, \tilde{h}]| < \varepsilon$ for all $x \in S$. For an arbitrary h set $\tilde{h} = (\delta/\|h\|)h$. Then $\|\tilde{h}\| = \delta$ and

$$|M[x, (\delta/\|h\|)h]| < \varepsilon, \qquad x \in S, \quad h \in X. \tag{5.3.3}$$

Since

$$M[x, (\delta/\|h\|)h] = (\delta/\|h\|)M[x,h], \tag{5.3.4}$$

it follows from (5.3.3) and (5.3.4) that

$$|M[x,h]| < (\varepsilon/\delta)\|h\|, \qquad x \in S, \quad h \in X.$$

The proof is complete with $K(r) = \varepsilon/\delta$.

Consider the abstract Cauchy problem

$$u' = f(t, u),$$ (5.3.5)

$$u(t_0) = u_0,$$ (5.3.6)

and the scalar initial value problem

$$r' = g(t, r),$$ (5.3.7)

$$r(t_0) = r_0,$$ (5.3.8)

where $f: R_+ \times X \to X$ and $g: R_+ \times R_+ \to R$.

In this section, we shall assume that f and g are smooth enough to guarantee the existence (not uniqueness) of solutions of (5.3.5) and (5.3.7) for all $t \in R_+$. Actually, local existence would suffice and our results can be easily restated to hold locally. Of course the mere continuity of f will not suffice even for local existence of solutions of (5.3.5). For existence theorems the reader is to refer to Section 5.2.

A solution of (5.3.5) and (5.3.6) will be denoted by $u(t, t_0, u_0)$. The maximal solution of (5.3.7) and (5.3.8) will be denoted by $r(t, t_0, r_0)$.

We shall also assume, in this section, the existence of an admissible functional Φ in X satisfying the properties (i)–(iii) of Definition 5.3.1.

For easy reference we state the following hypotheses.

Hypothesis 1:

$$M[x-y, f(t, x)-f(t, y)] \leqslant g[t, \Phi(x-y)]$$ (5.3.9)

for $t \in R_+$ and all $x, y \in X$.

Hypothesis 2: The function $f(t, x)$ has a continuous Fréchet derivative $f_x(t, x)$ with respect to x and

$$M[h, f_x(t, z)h] \leqslant g[t, \Phi(h)], \qquad t \geqslant 0, \quad h \in X$$ (5.3.10)

and all z in any sphere $S(x_0, r)$.

As we shall prove in Lemma 5.3.6, Hypothesis 2 implies Hypothesis 1. First we shall prove that under Hypothesis 1 the system (5.3.5) and (5.3.6) has a unique solution $u(t, t_0, u_0)$ which depends continuously on the initial conditions (t_0, u_0) provided that the scalar initial value problem (5.3.7) and (5.3.8) has these properties.

We need the following lemma. The symbol $D_+ r(t)$ denotes the lower right-Dini derivative of the function $r(t)$.

LEMMA 5.3.2.   For any differentiable function $x: R_+ \to X$ the following inequality holds:

$$D_+ \Phi[x(t)] \leqslant M[x(t), x'(t)], \qquad t \in R_+ . \qquad (5.3.11)$$

Proof: From the definition of $D_+ \Phi(t)$, the admissibility of Φ and lemma 5.3.1 we obtain

$$
\begin{aligned}
D_+ \Phi[x(t)] &= \liminf_{h \to 0_+} h^{-1} [\Phi(x(t+h)) - \Phi(x(t))] \\
&= \liminf_{h \to 0_+} h^{-1} (\Phi[x(t) + hx'(t) + O(h^2)] - \Phi[x(t)]) \\
&\leqslant \liminf_{h \to 0_+} [h^{-1} (M[x(t), hx'(t)] + O(h^2)) + O(\|hx'(t) + O(h^2)\|)] \\
&\leqslant M[x(t), x'(t)] + \liminf_{h \to 0_+} M[x(t), O(h)] \\
&= M[x(t), x'(t)].
\end{aligned}
$$

The proof is complete.

The following lemma is used to prove uniqueness.

LEMMA 5.3.3.   Let Hypothesis 1 be satisfied. Assume that $\Phi(u_0 - v_0) \leqslant r_0$. Then

$$\Phi[u(t, t_0, u_0) - v(t, t_0, v_0)] \leqslant r(t, t_0, r_0), \qquad t \geqslant t_0 . \qquad (5.3.12)$$

Proof: Let $u(t) = u(t, t_0, u_0)$ and $v(t) = v(t, t_0, v_0)$ be solutions of (5.3.5) through (t_0, u_0) and (t_0, v_0) respectively and $r(t, t_0, r_0)$ be the maximal solution of (5.3.7) and (5.3.8). Define $z(t) = u(t) - v(t)$. From Lemma 5.3.2 and Hypothesis 1 we obtain

$$
\begin{aligned}
D_+ \Phi(z(t)) &\leqslant M[z(t), z'(t)] \\
&= M[u(t) - v(t), f[t, u(t)] - f[t, v(t)]] \\
&\leqslant g(t, \Phi(z(t))), \qquad t \geqslant t_0 .
\end{aligned}
$$

Also

$$\Phi[z(t_0)] = \Phi(u_0 - v_0) \leqslant r_0 .$$

From these inequalities and [42] the estimate (5.3.12) follows.

THEOREM 5.3.1.   Let Hypothesis 1 be satisfied. Assume that $r(t, t_0, 0) \equiv 0$. Then the system (5.3.5) and (5.3.6) has a unique solution.

Proof: Let $u_1(t) = u_1(t, t_0, u_0)$ and $u_2(t) = u_2(t, t_0, u_0)$ be two solutions of (5.3.5) and (5.3.6). It follows from (5.3.12) that

$$\Phi[u_1(t) - u_2(t)] \leqslant r(t, t_0, r_0) \equiv 0.$$

Hence $u_1(t) \equiv u_2(t)$, for $t \geqslant t_0$ and the proof is complete.

REMARK 5.3.1. As we mentioned in Example 5.3.1 (iii) the functional $\Phi(x) = (x, x)$ is admissible in any Hilbert space H with $M[x, h] = 2\,\text{Re}(x, h)$. In this case Hypothesis 1 with $g(t, r) \equiv 2Mr$ reduces to the monotonicity condition on f that was used to prove existence and uniqueness of solutions of (5.3.5) and (5.3.6) in Corollary 5.2.1.

Next, we prove the continuous dependence of solutions of (5.3.5) and (5.3.6) with respect to the initial conditions.

THEOREM 5.3.2. Let Hypothesis 1 be satisfied. Assume that the maximal solution of (5.3.7) and (5.3.8) depends continuously on (t_0, r_0) for each $(t_0, r_0) \in R_+ \times R_+$ and $r(t, t_0, 0) \equiv 0$, for $t \geqslant t_0$. Then $u(t, t_0, u_0)$ depends continuously on (t_0, u_0).

Proof: Let $u_1(t) = u(t, t_1, u_1)$ and $u_2(t) = u(t, t_2, u_2)$ be solutions of (5.3.5) through (t_1, u_1) and (t_2, u_2), respectively. Let $t_2 \geqslant t_1$. Define

$$z(t) = u_1(t) - u_2(t).$$

From Lemma 5.3.2 and Hypothesis 1 we obtain

$$
\begin{aligned}
D_+ \Phi(z(t)) &\leqslant M[z(t), z'(t)] \\
&= M[z(t), f[t, u_1(t)] - f[t, u_2(t)]] \\
&\leqslant g(t, \Phi[z(t)]), \qquad t \geqslant t_1.
\end{aligned}
\tag{5.3.13}
$$

Also

$$\Phi[z(t_1)] = \Phi[u_1 - u(t_1, t_2, u_2)]. \tag{5.3.14}$$

From (5.3.13) and (5.3.14), it follows that

$$\Phi[z(t)] \leqslant r(t, t_1, \Phi[u_1 - u(t_1, t_2, u_2)]). \tag{5.3.15}$$

Since $\Phi(x)$ is continuous in x, $u(t, t_2, u_2)$ is continuous in t and $r(t, t_1, r_1)$ is by hypothesis continuous in (t_1, r_1), it follows from (5.3.15) that

$$
\begin{aligned}
\lim_{\substack{t_1 \to t_2 \\ u_1 \to u_2}} \Phi[z(t)] &\leqslant r[t, t_2, \Phi(u_2 - u_2)] \\
&= r(t, t_2, 0) \equiv 0.
\end{aligned}
$$

Hence, from the definition of Φ

$$\lim_{\substack{t_1 \to t_2 \\ u_1 \to u_2}} z(t) = 0$$

and the proof is complete.

Now we shall prove that under Hypothesis 2 and $r(t, t_0, 0) \equiv 0$ the solutions $u(t, t_0, u_0)$ of (5.3.5) and (5.3.6) are continuously differentiable with respect to initial conditions (t_0, u_0) and the Fréchet derivatives $(\partial/\partial u_0) u(t, t_0, u_0)$ and $(\partial/\partial t_0) u(t, t_0, u_0)$ exist and satisfy the equation of variation of (5.3.5) along the solution $u(t, t_0, u_0)$. From the existence and continuity of $f_x(t, x)$ and from the mean value theorem for Fréchet differentiable functions, it follows that $f(t, x)$ is locally Lipschitzian in x, and consequently the local existence of solutions of (5.3.5) and (5.3.6) is secured. The following lemmas are needed.

LEMMA 5.3.4.   Let $f \in C[R_+ \times S(x_0, r), X]$ and let $f_x(t, x)$ exist and be continuous for $x \in S(x_0, r)$. Then for $x_1, x_2 \in S(x_0, r)$ and $t \geqslant 0$

$$f(t, x_1) - f(t, x_2) = \int_0^1 f_x(t, sx_1 + (1-s)x_2)(x_1 - x_2) \, ds. \quad (5.3.16)$$

Proof: Define

$$F(s) = f[t, sx_1 + (1-s)x_2], \qquad 0 \leqslant s \leqslant 1.$$

The convexity of $S(x_0, r)$ implies that $F(s)$ is well defined. Using the chain rule for Fréchet derivatives, we obtain

$$F'(s) = f_x[t, sx_1 + (1-s)x_2](x_1 - x_2). \quad (5.3.17)$$

Since $F(1) = f(t, x_1)$ and $F(0) = f(t, x_2)$ the result follows by integrating (5.3.17) with respect to s from 0 to 1.

LEMMA 5.3.5.   Let Hypothesis 2 be satisfied. Then

$$M\left[h, \int_0^1 f_x[t, sx_1 + (1-s)x_2] h \, ds\right]$$
$$\leqslant g[t, \Phi(h)], \qquad t \geqslant 0, \qquad h, x_1, x_2 \in s(x_0, r). \quad (5.3.18)$$

Proof: Let $\pi: 0 = s_0 < s_1 < \cdots < s_n = 1$ be any partition of $[0, 1]$. From the definition of Riemann integral for continuous abstract functions it follows that

$$\int_0^1 f_x[t, sx_1 + (1-s)x_2] h \, ds = \lim_{n \to \infty} \left[\sum_{i=0}^{n-1} f_x[t, \tau_i x_1 + (1-\tau_i)x_2] h \, \Delta s_i \right]$$

(5.3.19)

where $\Delta s_i = s_{i+1} - s_i$ and $s_i \leqslant \tau_i \leqslant s_{i+1}$, for $i = 0, 1, ..., n-1$.

From the continuity of $M[x, h]$ in h (uniformly with respect to $x \in S(x_0, r)$), (5.3.19), and Hypothesis 2 it follows that

$$M\left[h, \int_0^1 f_x[t, sx_1 + (1-s)x_2] h \, ds \right]$$

$$= \lim_{n \to \infty} M\left[h, \sum_{i=0}^{n-1} f_x(t, \tau_i x_1 + (1-\tau_i)x_2) h \Delta s_i \right]$$

$$\leqslant \lim_{n \to \infty} \sum_{i=0}^{n-1} \Delta s_i \, M[h, f_x(t, \tau_i x_1 + (1-\tau_i)x_2)h]$$

$$\leqslant g(t, \Phi(h)) \lim_{n \to \infty} \sum_{i=0}^{n-1} \Delta s_i = g(t, \Phi(h)).$$

The proof is complete.

LEMMA 5.3.6. Hypothesis 2 implies Hypothesis 1.

Proof: In view of Lemmas (5.3.4) and (5.3.5) we have

$$M[x-y, f(t,x) - f(t,y)] = M\left[x-y, \int_0^1 f_x[t, sx + (1-s)y](x-y) \, ds \right]$$

$$\leqslant g[t, \Phi(x-y)].$$

The proof is complete.

COROLLARY 5.3.1. Let Hypothesis 2 be satisfied. Assume that the maximal solution of (5.3.7) and (5.3.8) depends continuously on (t_0, r_0) for each $(t_0, r_0) \in R_+ \times R_+$ and $r(t, t_0, 0) \equiv 0$ for $t \geqslant t_0$. Then the solutions of (5.3.5) and (5.3.6) exist locally, are unique, and depend continuously on initial conditions.

THEOREM 5.3.3. Let Hypothesis 2 be satisfied. Assume that the maximal solution of (5.3.7) through any point $(t_0, 0)$ is identically zero for $t \geqslant t_0$.

Then

(a) The Fréchet derivative $(\partial/\partial u_0)u(t, t_0, u_0) \equiv U(t, t_0, u_0)$ exists and satisfies the operator equation

$$U' = f_u[t, u(t, t_0, u_0)]\, U, \qquad t \geq t_0, \tag{5.3.20}$$

$$U(t_0) = I; \tag{5.3.21}$$

(b) the Fréchet derivative $(\partial/\partial t_0)u(t, t_0, u_0) \equiv V(t, t_0, u_0)$ exists and satisfies

$$V' = f_u(t, u(t, t_0, u_0))V, \qquad t \geq t_0, \tag{5.3.22}$$

$$V(t_0) = -f(t_0, u_0). \tag{5.3.23}$$

Furthermore

$$V(t, t_0, u_0) = -U(t, t_0, u_0)f(t_0, u_0). \tag{5.3.24}$$

Proof: (a) Since $f_u(t, \cdot) \in C[J \times S(u_0, r), B(X)]$, (5.3.20) and (5.3.21) has a unique solution which we denote by $U(t)$.

Define the function

$$z(t) = u(t, t_0, u_0 + h) - u(t, t_0, u_0) - U(t)h,$$

$$t \geq t_0, \quad u_0, u_0 + h \in S(u_0, r).$$

Then

$$D_+ \Phi\big(z(t)/\|h\|\big)$$

$$\leq M[z(t)/\|h\|, z'(t)/\|h\|]$$

$$= M[z(t)/\|h\|, f[t, u(t, t_0, u_0 + h)] - f[t, u(t, t_0, u_0)]/\|h\|$$

$$- f_u[t, u(t, t_0, u_0 + h)]\, U(t)h/\|h\|]. \tag{5.3.25}$$

From the Fréchet differentiability of f with respect to $u \in S(u_0, r)$, we have

$$f[t, u(t, t_0, u_0 + h)] - f[t, u(t, t_0, u_0)]$$

$$= f_u[t, u(t, t_0, u_0)][u(t, t_0, u_0 + h) - u(t, t_0, u_0)]$$

$$+ O(\|u(t, t_0, u_0 + h) - u(t, t_0, u_0)\|)$$

$$= f_u[t, u(t, t_0, u_0)][z(t) + U(t)h]$$

$$+ O(\|u(t, t_0, u_0 + h) - u(t, t_0, u_0\|). \tag{5.3.26}$$

Set

$$\omega(h) = O(\|u(t, t_0, u_0 + h) - u(t, t_0, u_0)\|).$$

From (5.3.25) and (5.3.26) we obtain

$$D_+ \Phi(z(t)/\|h\|) \leqslant M[z(t)/\|h\|, f_u[t, u(t, t_0, u_0)] z(t)/\|h\| + \omega(h)/\|h\|]$$

$$\leqslant M[z(t)/\|h\|, f_u[t, u(t, t_0, u_0)] z(t)/\|h\|]$$

$$+ M[z(t)/\|h\|, \omega(h)/\|h\|]$$

$$\leqslant g[t, \Phi(z(t)/\|h\|)] + M[z(t)/\|h\|, \omega(h)/\|h\|]. \quad (5.3.27)$$

Next we shall prove that in any compact interval of t

$$\lim_{\|h\| \to 0} M[z(t)/\|h\|, \omega(h)/\|h\|] = 0. \quad (5.3.28)$$

Define

$$m(t) = \|u(t, t_0, u_0 + h) - u(t, t_0, u_0)\|. \quad (5.3.29)$$

It follows that

$$D_+ m(t) \leqslant \|u'(t, t_0, u_0 + h) - u'(t, t_0, u_0)\|$$

$$= \|f[t, u(t, t_0, u_0 + h)] - f[t, u(t, t_0, u_0)]\|.$$

In view of Lemma 5.3.4 and the continuity of f we obtain from (5.3.29)

$$D_+ m(t) \leqslant K_1 m(t) \quad (5.3.30)$$

where K_1 is a constant such that

$$\|f_u(t, z)\| \leqslant K_1$$

for t in a compact interval I around t_0 and z being in the line segment joining the solutions $x(t, t_0, x_0)$ and $x(t, t_0, x_0 + h)$. Also from (5.3.29)

$$m(t_0) = \|h\|. \quad (5.3.31)$$

By (5.3.30) and (5.3.31) we obtain

$$m(t) \leqslant \|h\| \exp K_1(t - t_0)$$

$$\leqslant K_2 \|h\|, \quad t \in I \quad (5.3.32)$$

where K_2 is a constant.

Let K_3 be a constant such that

$$\|U(t)\| \leqslant K_3, \quad t \in I. \quad (5.3.33)$$

From the definition of $z(t)$, (5.3.32), and (5.3.33) we get

$$\|z(t)\|/\|h\| \leqslant K_2 + K_3 \equiv K_4, \qquad t \in I, \quad \|h\| \text{ sufficiently small.} \qquad (5.3.34)$$

In view of Lemma 5.3.1 [which applies because of (5.3.34)] we have

$$M[z(t)/\|h\|, \omega(h)/\|h\|] \leqslant K\|\omega(h)\|/\|h\|. \qquad (5.3.35)$$

Finally, from the definition of $\omega(h)$ and (5.3.32) we obtain

$$\|\omega(h)\|/\|h\| \leqslant K_2 \|\omega(h)\|/m(t)$$

$$= K_2 \|\omega(h)\|/\|u(t, t_0, u_0 + h) - u(t, t_0, u_0)\| \to 0 \qquad \text{as} \quad h \to 0$$

and (5.3.28) has been established.

From (5.3.27) and (5.3.28) we obtain

$$D_+ \Phi(z(t)/\|h\|) \leqslant g[t, \Phi(z(t)/\|h\|)] + O(1). \qquad (5.3.36)$$

Also

$$\Phi(z(t_0)/\|h\|) = 0. \qquad (5.3.37)$$

In view of (5.3.36) and (5.3.37) it follows that

$$\lim_{\|h\| \to 0} \Phi(z(t)/\|h\|) = r(t, t_0, 0) \equiv 0.$$

Hence, from Definition 5.3.1

$$\lim_{\|h\| \to 0} z(t)/\|h\| = 0,$$

which proves that the Fréchet derivative $(\partial/\partial u_0)u(t, t_0, u_0)$ exists and it is equal to $U(t)$. The proof of (a) is complete.

(b)   Let $U(t)$ be as in (a) the solution of (5.3.20) and (5.3.21). Define the function

$$z(t) = u(t, t_0 + h, u_0) - u(t, t_0, u_0) + U(t)f(t_0, u_0)h$$

Then as in (a)

$$D_+ \Phi(z(t)/h) \leqslant M[z(t)/h, z'(t)/h]$$

$$= M[z(t)/h, (f[t, u(t, t_0 + h, u_0)] - f[t, u(t, t_0, u_0)])/h$$

$$+ f_u[t, u(t, t_0, u_0)] U(t)f(t_0, u_0)]$$

$$= M[z(t)/h, f_u[t, u(t, t_0, u_0)] z(t)/h + \omega(h)/h]$$

$$\leqslant M[z(t)/h, f_u[t, u(t, t_0, u_0)](z(t)/h)] + M[z(t)/h, \omega(h)/h]$$

$$\leqslant g[t, \Phi(z(t)/h)] + O(1) \qquad \text{as} \quad h \to 0. \qquad (5.3.38)$$

Also

$$\Phi(z(t_0)/h) = \Phi([u(t_0, t_0+h, u_0) - u_0]/h + f(t_0, u_0))$$

$$= \Phi([u(t_0, t_0+h, u_0) - u(t_0+h, t_0+h, u_0)]/h + f(t_0, u_0))$$

$$= O(1) \quad \text{as} \quad h \to 0. \tag{5.3.39}$$

From (5.3.38), (5.3.39), and Lakshmikantham and Leela [42] it follows as in (a) that

$$\lim_{h \to 0} \Phi(z(t)/h) = r(t, t_0, 0) \equiv 0.$$

Hence

$$\lim_{h \to 0_+} z(t)/h = 0.$$

which proves (b). In addition

$$(\partial/\partial t_0) u(t, t_0, u_0) = -U(t) f(t_0, u_0)$$

$$= -(\partial/\partial u_0) u(t, t_0, u_0) f(t_0, u_0).$$

The proof is complete.

THEOREM 5.3.4. Under the hypotheses of Theorem 5.3.3 the following formula holds:

$$u(t, t_0, v_0) - u(t, t_0, u_0) = \int_0^1 U[t, t_0, u_0 + s(v_0 - u_0)](v_0 - u_0)\, ds. \tag{5.3.40}$$

Proof: From Theorem 5.3.3 and the chain rule for abstract functions, we have

$$(d/ds) u[t, t_0, u_0 + s(v_0 - u_0)] = U[t, t_0, u_0 + s(v_0 - u_0)](v_0 - u_0). \tag{5.3.41}$$

Integrating (5.3.41) from 0 to 1 with respect to s the desired result follows.

Now we shall establish the variation of constants formula with respect to (5.3.5) and (5.3.6) and its nonlinear perturbation

$$v' = f(t, v) + F(t, v), \tag{5.3.42}$$

$$v(t_0) = v_0 \tag{5.3.43}$$

where $f, F: R_+ \times X \to X$ are smooth enough to guarantee the existence of solutions of (5.3.42) and (5.3.43) for $t \geq t_0$. A solution of (5.3.42) and (5.3.43) is denoted by $v(t, t_0, v_0)$.

THEOREM 5.3.5.   Let $f, F \in C[R_+ \times X; X]$ and let f satisfy Hypothesis 2. Let $u(t, t_0, u_0)$ and $v(t, t_0, u_0)$ be solutions of (5.3.5) and (5.3.42) through (t_0, u_0), respectively. Then, for $t \geq t_0$

$$v(t, t_0, u_0) = u(t, t_0, u_0) + \int_{t_0}^{t} U[t, s, v(s, t_0, u_0)] F[s, v(s, t_0, u_0)] \, ds$$
$$(5.3.44)$$

where

$$U(t, t_0, u_0) = (\partial/\partial u_0) u(t, t_0, u_0).$$

Proof: Write $v(t) = v(t, t_0, u_0)$. Then, in view of Theorem 5.3.3

$$\begin{aligned}
(d/ds) u[t, s, v(s)] &= (\partial/\partial s) u[t, s, v(s)] + (\partial/\partial v) u[t, s, v(s)] v'(s) \\
&= -U[t, s, v(s)] f(s, v(s)) \\
&\quad + U[t, s, v(s)] (f[s, v(s)] + F[s, v(s)]) \\
&= U[t, s, v(s)] F[s, v(s)].
\end{aligned}$$
$$(5.3.45)$$

Since the right-hand side of (5.3.45) is continuous, we can integrate from t_0 to t, obtaining the variation of constants formula (5.3.44).

5.4. Stability and Asymptotic Behavior

Let us consider the abstract differential equation

$$u' = f(t, u) \qquad \text{and} \qquad u(t_0) = u_0 \tag{5.4.1}$$

and its perturbation

$$v' = f(t, v) + F(t, v) \qquad \text{and} \qquad v(t_0) = v_0 \tag{5.4.2}$$

where $f, F \in C[R_+ \times S(p), X]$, $S(p)$ being the sphere $\{u \in X: \|u\| < p\}$ in the Banach space X. We assume that the functions f and F are smooth enough to ensure the existence of solutions $u(t, t_0, u_0)$ and $v(t, t_0, v_0)$ of (5.4.1) and (5.4.2), respectively, on $[t_0, \infty)$. When $f(t, 0) \equiv 0$, (5.4.1) has the trivial solution. In this case we have the following.

DEFINITION 5.4.1.   The trivial solution of (5.4.1) is said to be

(i)   *stable* if for every $\varepsilon > 0$ and $t_0 \in R_+$, there exists a $\delta > 0$ such that $\|u_0\| < \delta$ implies $\|u(t, t_0, u_0)\| < \varepsilon$ for all $t \geq t_0$;

(ii)   *asymptotically stable* if it is stable and there exists a $\delta_0 > 0$ such that $\|u_0\| < \delta_0$ implies $\lim_{t \to \infty} u(t, t_0, u_0) = 0$;

(iii)   *uniformly stable in variation* if for every $\varepsilon > 0$ and $t_0 \in R_+$, there exists an $M(\varepsilon) > 0$ such that $\|u_0\| < \varepsilon$ implies $\|U(t, t_0, u_0)\| < M(\varepsilon)$ for all $t \geqslant t_0$, where $U(t, t_0, u_0)$ is the solution of the variational equation (5.3.20) and (5.3.21).

DEFINITION 5.4.2.   Let $A \in B(X)$. The *logarithmic norm* of the operator A is defined by

$$\mu(A) \equiv \lim_{h \to 0_+} (\|I + hA\| - 1)/h. \tag{5.4.3}$$

PROBLEM 5.4.1.   Prove that the limit in (5.4.3) exists and satisfies the properties

(i) $|\mu(A)| \leqslant \|A\|$;

(ii) $\mu(\alpha A) = \alpha \mu(A), \qquad \alpha \geqslant 0$;

(iii) $\mu(A + B) \leqslant \mu(A) + \mu(B)$;

(iv) $|\mu(A) - \mu(B)| \leqslant \|A - B\|$.

[Hint: The right Gateaux derivative of $\|x\|$ exists in any Banach space.]

The following lemma which is interesting in itself is needed for further considerations.

LEMMA 5.4.1.   Let $A(t) \in B(X)$ for each $t \in R_+$ and suppose that $u(t)$ is the solution of

$$u' = A(t)u \qquad \text{and} \qquad u(t_0) = u_0.$$

Then

$$\|u(t)\| \leqslant \|u_0\| \exp\left(\int_{t_0}^{t} \mu[A(s)]\, ds\right), \qquad t \geqslant t_0. \tag{5.4.4}$$

Proof: Define $m(t) = \|u(t)\|$. Then, for small $h > 0$,

$$m(t+h) - m(t) \leqslant \|u(t) + hA(t)u(t)\| - \|u(t)\| + \varepsilon(h)$$

$$\leqslant (\|I + hA(t)\| - 1)m(t) + \varepsilon(h)$$

where $\varepsilon(h)/h \to 0$ as $h \to 0_+$. Hence

$$D_+ m(t) \leqslant \mu[A(t)]m(t) \qquad \text{and} \qquad m(t_0) = \|u_0\|$$

from which the estimate (5.4.4) follows.

LEMMA 5.4.2.   Let the hypotheses of Theorem 5.3.3 hold. Suppose further that there exists a function $\alpha \in C[R_+, R]$ such that

$$\mu[f_u(t, u)] \leqslant \alpha(t), \qquad (t, u) \in R_+ \times S(p). \tag{5.4.5}$$

Then for $u_0, v_0 \in S(p)$ we have the estimates

$$\|u(t, t_0, v_0) - u(t, t_0, u_0)\| \leqslant \|v_0 - u_0\| \exp\left[\int_{t_0}^{t} \alpha(s)\, ds\right], \qquad t \geqslant t_0 \tag{5.4.6}$$

and

$$\|v(t, t_0, v_0) - u(t, t_0, u_0)\|$$
$$\leqslant \|v_0 - u_0\| \exp\left[\int_{t_0}^{t} \alpha(s)\, ds\right]$$
$$+ \int_{t_0}^{t} \exp\left[\int_{s}^{t} \alpha(\xi)\, d\xi\right] \|F[s, v(s, t_0, v_0)]\|\, ds, \qquad t \geqslant t_0. \tag{5.4.7}$$

Proof: From Theorem 5.3.4 we have

$$u(t, t_0, v_0) - u(t, t_0, u_0) = \int_{0}^{1} U[t, t_0, u_0 + s(v_0 - u_0)](v_0 - u_0)\, ds. \tag{5.4.8}$$

By virtue of (5.4.5) and Lemma 5.4.1 it follows that

$$\max_{0 \leqslant s \leqslant 1} \|U[t, t_0, u_0 + s(v_0 - u_0)]\| \leqslant \exp \int_{t_0}^{t} \alpha(s)\, ds.$$

This and (5.4.8) yield (5.4.6).

Next, from Theorem 5.3.5 we have

$$v(t, t_0, v_0) - u(t, t_0, u_0)$$
$$= u(t, t_0, v_0) - u(t, t_0, u_0)$$
$$+ \int_{t_0}^{t} U[t, s, v(s, t_0, v_0)]\, F[s, v(s, t_0, v_0)]\, ds. \tag{5.4.9}$$

Again, from Lemma 5.4.1

$$\max_{t_0 \leqslant s \leqslant t} \|U[t, s, v(s, t_0, v_0)]\| \leqslant \exp \int_{s}^{t} \alpha(\xi)\, d\xi.$$

This, together with (5.4.6) and (5.4.9) yields the estimate (5.4.7). The proof is complete.

THEOREM 5.4.1. Assume that

(i) the hypotheses of Theorem 5.3.3 hold;

(ii) $\qquad\qquad f(t, 0) \equiv 0, \qquad t \in R_+$;

(iii) the condition (5.4.5) holds with

$$\sigma \equiv \lim_{t \to \infty} \sup (t - t_0)^{-1} \int_{t_0}^{t} \alpha(s)\, ds < 0. \qquad (5.4.10)$$

Then the trivial solution of (5.4.1) is asymptotically stable.

Proof: The assumption (5.4.10) implies that

$$\int_{t_0}^{t} \alpha(s)\, ds \leq (\sigma/2)(t - t_0), \qquad t \text{ sufficiently large.}$$

Therefore

$$\lim_{t \to \infty} \exp\left[\int_{t_0}^{t} \alpha(s)\, ds\right] = 0. \qquad (5.4.11)$$

If we take $v_0 = 0$ in (5.4.6) we obtain

$$\|u(t, t_0, u_0)\| \leq \|u_0\| \exp\left[\int_{t_0}^{t} \alpha(s)\, ds\right].$$

This and (5.4.11) yield the desired conclusion.

THEOREM 5.4.2. Assume that

(i) the hypotheses of Theorem 5.3.3 hold;

(ii) $\qquad f(t, 0) \equiv F(t, 0) \equiv 0, \qquad t \in R_+$;

(iii) $\qquad \sigma < 0$;

(iv) $\qquad \|F(t, v)\| = O(\|v\|) \qquad$ as $\quad v \to 0 \quad$ uniformly in t.

Then, the trivial solution of (5.4.2) is asymptotically stable.

Proof: Let $\varepsilon > 0$ and sufficiently small. Then Hypothesis (iii) implies that

$$\lim_{t \to \infty} \exp\left[\varepsilon(t - t_0) + \int_{t_0}^{t} \alpha(s)\, ds\right] = 0. \qquad (5.4.12)$$

Hence, there exists a positive constant K such that

$$\exp\left[\varepsilon(t - t_0) + \int_{t_0}^{t} \alpha(s)\, ds\right] \leq K, \qquad t \geq t_0. \qquad (5.4.13)$$

With the foregoing ε and because of Hypothesis (iv) there exists a $\delta > 0$ such that $\|v\| < \delta$ implies $\|F(t,v)\| < \varepsilon \|v\|$. Now for $\|v_0\| < \delta/K$ and from (5.4.7) with $u_0 = 0$ we obtain

$$\|v(t,t_0,v_0)\| \leqslant \|v_0\| \exp\left[\int_{t_0}^{t} \alpha(s)\,ds\right]$$

$$+ \varepsilon \int_{t_0}^{t} \exp\left[\int_{s}^{t} \alpha(\xi)\,d\xi\right] \|v(s,t_0,v_0)\|\,ds \qquad (5.4.14)$$

as long as $\|v(t,t_0,v_0)\| < \delta$. Multiplying both sides of (5.4.14) by $\exp[-\int_{t_0}^{t} \alpha(s)\,ds]$ and applying Gronwall's inequality we get

$$\|v(t,t_0,v_0)\| \leqslant \|v_0\|\left[\varepsilon(t-t_0) + \int_{t_0}^{t} \alpha(s)\,ds\right] \qquad (5.4.15)$$

as long as $\|v(t,t_0,v_0)\| < \delta$. Now (5.4.15) shows that $\|v(t,t_0,v_0)\| < \delta$ for all $t \geqslant t_0$. Otherwise there exists a T such that $\|v(T,t_0,v_0)\| = \delta$ and $\|v(t,t_0,v_0)\| \leqslant \delta$ for $t_0 \leqslant t \leqslant T$. Then from (5.4.15) and (5.4.13), we get the contradiction

$$\delta < (\delta/K)K = \delta$$

which proves our claim. Thus (5.4.15) holds for all $t \geqslant t_0$ and this together with (5.4.12) yields the desired conclusion.

THEOREM 5.4.3.   Assume that

(i)   the hypotheses of Theorem 5.3.3 are satisfied;

(ii) $f(t,0) \equiv 0, \qquad t \in R_+$;

(iii)   the trivial solution of (5.4.1) is uniformly stable in variation;

(iv)   given $\alpha > 0$ there exists a function $\lambda_\alpha \in L_1[0,\infty)$ such that $\|F(t,u)\| \leqslant \lambda_\alpha(t)$ for $\|u\| \leqslant \alpha$.

Then for every $\varepsilon > 0$ there exists positive numbers $\delta = \delta(\varepsilon)$ and $T = T(\varepsilon)$ such that $\|u_0\| < \delta$ implies $\|v(t,t_0,v_0)\| < \varepsilon$ for $t \geqslant t_0 \geqslant T$.

Proof: Let $\varepsilon > 0$ be given. Choose δ and T such that $\delta < \varepsilon$, $2M(\varepsilon)\delta < \varepsilon$ and $\int_T^\infty \lambda_\varepsilon(s)\,ds < \varepsilon/2M(\varepsilon)$. Assume that $\|u_0\| < \delta$ and $t_0 \geqslant T$. Using (5.3.40) with $v_0 = 0$ and Hypothesis (iii) we get

$$\|u(t,t_0,u_0)\| \leqslant \|u_0\| M(\varepsilon) \leqslant \varepsilon/2, \qquad t \geqslant t_0.$$

We claim that

$$\|v(t,t_0,u_0)\| < \varepsilon, \qquad t \geqslant t_0.$$

If this is not true, let $t_1 > t_0$ be such that $\|v(t_1, t_0, u_0)\| = \varepsilon$ and $\|v(t, t_0, u_0)\| \leqslant \varepsilon$ for $t_0 \leqslant t \leqslant t_1$. Then from (5.3.44)

$$\varepsilon \leqslant \varepsilon/2 + M(\varepsilon) \int_{t_0}^{t} \lambda_\varepsilon(s) \, ds < \varepsilon.$$

This contradiction proves the theorem.

5.5. Chaplygin's Method

In this section we shall employ Chaplygin's method to approximate the solution of the differential system

$$u' = f(t, u) \tag{5.5.1}$$

$$u(0) = u_0 \tag{5.5.2}$$

by a sequence of functions $u_n(t)$ which satisfy the linear systems

$$u'_{n+1}(t) = f[t, u_n(t)] + f_u[t, u_n(t)][u_{n+1}(t) - u_n(t)] \tag{5.5.3}$$

and

$$u_{n+1}(0) = u_0, \qquad n = 0, 1, 2, \ldots. \tag{5.5.4}$$

The right-hand side of (5.5.3) is a linear approximation of $f(t, u)$. This is the analog of Newton's method applied to numerical equations and it was used by Chaplygin for ordinary differential equations in R^n. Here $f \in C[[0, a] \times X, X]$ where $a > 0$ and X is a Banach space. We shall assume that for each t the function $f(t, u)$ is Fréchet differentiable in u with F-derivative at the point $v \in X$ denoted by $f_u(t, v)$. Furthermore, we shall assume that $f_u(t, v)$ is strongly continuous in (t, v) and

$$\|f_u(t, v) - f_u(t, w)\| \leqslant g(t, \|v - w\|), \qquad v, w \in X \quad t \in [0, a] \tag{5.5.5}$$

where $g \in C[[0, a] \times R_+, R_+]$ and $g(t, r)$ is nondecreasing in r.

If $u_0(t)$ is a continuous function on $[0, a)$ with $u_0(0) = u_0$, then in view of our hypotheses on f the system (5.5.3) and (5.5.4) with $n = 0$ has a unique solution $u_1(t)$ which exists on $[0, a]$. In this way, one can construct a sequence of functions $\{u_n(t)\}$, $n = 0, 1, 2, \ldots$ which satisfy (5.5.3) and (5.5.4) on $[0, a]$. Now we can prove the following.

THEOREM 5.5.1. Let $f(t, u)$ satisfy (5.5.5) and suppose that $\|u_n(t)\| \leqslant M < \infty$ for $0 \leqslant t \leqslant a$ and all n. Then $\{u_n(t)\}$ is uniformly convergent on $[0, a]$ to the solution $u(t)$ of the system (5.5.1) and (5.5.2).

Proof: It follows from the continuity of $f_u(t, 0)$ and from the principle of uniform boundedness that

$$\sup_{[0, a]} \|f_u(t, 0)\| = N < \infty.$$

The function $z_n(t) = u_{n+1}(t) - u_n(t)$ satisfies the system

$$z_n'(t) = f[t, u_n(t)] + f_u[t, u_n(t)] z_n(t)$$
$$- f[t, u_{n-1}(t)] - f_u[t, u_{n-1}(t)] z_{n-1}(t) \qquad (5.5.6)$$

and

$$z_n(0) = 0. \qquad (5.5.7)$$

We shall now prove the following estimates:

$$I_1 \equiv \|f[t, u_n(t)] - f[t, u_{n-1}(t)] - f_u[t, u_{n-1}(t)] z_{n-1}(t)\|$$
$$\leqslant g(t, \|z_{n-1}(t)\|) \|z_{n-1}(t)\| \qquad (5.5.8)$$

and

$$I_2 \equiv \|f_u[t, u_n(t)] z_n(t)\| \leqslant K \|z_n(t)\| \qquad (5.5.9)$$

where

$$K = N + \max_{[0, a]} g(t, M) < \infty. \qquad (5.5.10)$$

To prove (5.5.8) we shall employ the mean value theorem. Let $\phi \in X^*$ be a bounded real linear functional on X such that $\|\phi\| = 1$ and

$$I_1 = \phi(f[t, u_n(t)] - f[t, u_{n-1}(t)] - f_u[t, u_{n-1}(t)] z_{n-1}(t)).$$

The real function

$$F(\xi) \equiv \phi[f(t, u_{n-1}(t) + \xi[u_n(t) - u_{n-1}(t)])]$$

is differentiable in ξ and

$$F'(\xi) = \phi[f_u(t, u_{n-1}(t) + \xi[u_n(t) - u_{n-1}(t)]) z_{n-1}(t)].$$

Thus there exists a $\tau \in (0, 1)$ such that

$$F(1) - F(0) = F'(\tau),$$

that is

$$\phi(f[t, u_n(t)] - f[t, u_{n-1}(t)]) = \phi[f_u[t, u_{n-1}(t) + \tau z_{n-1}(t)] z_{n-1}(t)].$$

Hence

$$I_1 = \phi\left[f_u[t, u_{n-1}(t) + \tau z_{n-1}(t)] z_{n-1}(t) - f_u[t, u_{n-1}(t)] z_{n-1}(t)\right]$$
$$\leqslant \|f_u[t, u_{n-1}(t) + \tau z_{n-1}(t)] z_{n-1}(t) - f_u[t, u_{n-1}(t)] z_{n-1}(t)\|$$
$$\leqslant g(t, \tau \|z_{n-1}(t)\|) \|z_{n-1}(t)\|,$$

which proves (5.5.8). Next

$$\|f_u[t, u_n(t)]\| \leqslant \|f_u[t, u_n(t)] - f_u(t, 0)\| + \|f_u(t, 0)\|$$
$$\leqslant g(t, \|u_n(t)\|) + \max_{[0, a]} \|f_u(t, 0)\|$$
$$\leqslant g(t, M) + N$$
$$\leqslant \max_{[0, a]} g(t, M) + N = K,$$

which proves (5.5.9).

From (5.5.6)–(5.5.9) it follows that

$$\|z_n(t)\| \leqslant \int_0^t \exp[K(t-s)] g(s, \|z_{n-1}(s)\|) \|z_{n-1}(s)\| \, ds.$$

On the other hand, $\|z_n(t)\| \leqslant 2M$. Hence

$$\|z_n(t)\| \leqslant 2M \int_0^t \exp[K(t-s)] g(s, 2M) \, ds. \tag{5.5.11}$$

From (5.5.11) one establishes by induction that

$$\|z_n(t)\| \leqslant 2M(Ft)^{n-1}/(n-1)!, \qquad n = 1, 2, \ldots$$

where $F = R \exp Ra$ and $R = \max[K; \max_{[0, a]} g(t, 2M)]$.

From this estimate and the completeness of X we conclude that the sequence $\{u_n(t)\}$ converges uniformly on $[0, a]$ to a limit $u(t)$. In view of (5.5.3) and (5.5.4) we obtain

$$u_{n+1}(t) = u_0 + \int_0^t (f[s, u_n(s)] + f_u[s, u_n(s)] [u_{n+1}(s) - u_n(s)]) \, ds$$

and passing to the limit

$$u(t) = u_0 + \int_0^t f[s, u(s)] \, ds.$$

Hence $u(t)$ is a solution of the system (5.5.1) and (5.5.2). Since $f_u(t, u)$ is continuous the function $f(t, u)$ is locally Lipschitzian and therefore the system (5.5.1) and (5.5.2) has a unique solution. The proof is therefore complete.

The following theorem gives a satisfactory way to approximate the solution $u(t)$ of (5.5.1) and (5.5.2).

THEOREM 5.5.2.   Assume all the hypotheses of Theorem 5.5.1 and in addition

$$\|u_1(t) - u(t)\| \le w_1(t), \qquad 0 \le t \le a.$$

Define

$$w_{n+1}(t) = \int_0^t \exp[K(t-s)] g[s, w_n(s)]\, ds, \qquad n = 1, 2, \ldots$$

where $K = \sup_{[0,a]} \|f_u(t,0)\| + \max_{[0,a]} g(t, M)$.
  Then

$$\|u_n(t) - u(t)\| \le w_n(t), \qquad n = 1, 2, \ldots.$$

Proof: Define

$$m_n(t) = \|u_n(t) - u(t)\|, \quad n = 1, 2, \ldots.$$

Observe that

$$[u_n(t) - u(t)]' = f[t, u_{n-1}(t)] + f_u[t, u_{n-1}(t)][u_n(t) - u_{n-1}(t)] - f[t, u(t)]$$

$$= f_u[t, u_{n-1}(t)][u_n(t) - u(t)]$$

$$+ f[t, u_{n-1}(t)] - f[t, u(t)] + f_u[t, u_{n-1}(t)][u(t) - u_{n-1}(t)].$$

Replacing $u_n(t)$ by $u(t)$ in (5.5.8) and (5.5.9) we obtain

$$D_- m_n(t) \le K m_n(t) + g[t, m_{n-1}(t)] m_{n-1}(t)$$

and

$$m_n(0) = 0.$$

Hence

$$m_n(t) \le \int_0^t \exp[K(t-s)] g[s, m_{n-1}(s)] m_{n-1}(s)\, ds.$$

The desired result now follows by induction and the nondecreasing character of $g(t, u)$ in u.

PROBLEM 5.5.1.   Develop Chaplygin's method for the system

$$u' = Au + f(t, u), \qquad 0 \leqslant t \leqslant a, \qquad u(0) = u_0 \in D(A)$$

where A is the infinitesimal generator of a contraction semigroup and f is as before.

[Hint: Consider the sequence $\{u_n(t)\}$ satisfying

$$u'_{n+1}(t) = Au_{n+1}(t) + f(t, u_n(t)) + f_u(t, u_n(t))[u_{n+1}(t) - u_n(t)],$$

$$0 \leqslant t \leqslant a,$$

$$u'_{n+1}(0) = u_0, \qquad u_0(t) \equiv u_0, \qquad 0 \leqslant t \leqslant a.]$$

5.6. Global Existence and Asymptotic Equilibrium

The last counterexample presented in Section 5.1 shows how badly the solutions of an abstract Cauchy problem may behave in the case of an infinite dimensional Banach space. Although the function $f(t, u)$ is locally Lipschitzian in u and continuous for all points (t, u) we exhibited a solution $u(t)$ of (5.1.12) and (5.1.13) which exists on $(0, 1]$ and is bounded, but $u(t)$, contrary to the case in R^n, does not tend to a limit as $t \to \infty$. In this section we shall impose a condition on $f(t, u)$ which rules out such a behavior and which guarantees the global existence of solutions of the abstract Cauchy problem

$$du/dt = f(t, u), \qquad t \geqslant t_0, \tag{5.6.1}$$

and

$$u(t_0) = u_0 \tag{5.6.2}$$

where $f: J \times X \to X$ with $J = [t_0, \infty)$ and X is a Banach space.

We shall assume, without further mention, that $f(t, u)$ is smooth enough to assure local existence of solutions of (5.6.1) through any point in $J \times X$. For example, $f(t, x)$ may be locally Lipschitzian in x as in the counterexample or satisfy some monotonicity condition.

THEOREM 5.6.1.   Assume that

(i) $f \in C[J \times X, X]$ and for all $(t, u) \in J \times X$

$$\|f(t, u)\| \leqslant g(t, \|u\|); \tag{5.6.3}$$

(ii) $g \in C[J \times R_+, R_+]$ and $g(t,r)$ is nondecreasing in $r \geqslant 0$ for each $t \in J$, and the maximal solution $r(t, t_0, r_0)$ of the scalar initial value problem

$$r' = g(t,r) \tag{5.6.4}$$

and

$$r(t_0) = r_0 \tag{5.6.5}$$

exists throughout J.

Then the largest interval of existence of any solution $u(t, t_0, u_0)$ of (5.6.1) and (5.6.2) with $\|u_0\| \leqslant r_0$ is J. In addition if $r(t, t_0, r_0)$ is bounded on J then the (strong) $\lim_{t \to \infty} u(t, t_0, u_0)$ exists and is a (finite) vector in X.

Proof: Let $u(t) = u(t, t_0, u_0)$ be a solution of (5.6.1) and (5.6.2) with $\|u_0\| \leqslant r_0$ which exists on $[t_0, \beta)$ for $t_0 < \beta < \infty$ and such that the value of β cannot be increased (as in the counterexample). Define $m(t) = \|u(t)\|$ for $t_0 \leqslant t < \beta$. Then using (5.6.3) we obtain

$$D_+ m(t) \leqslant \|u'(t)\|$$

$$= \|f(t, u(t))\|$$

$$\leqslant g(t, m(t)), \qquad t_0 \leqslant t < \beta, \tag{5.6.6}$$

and

$$m(t_0) = \|u_0\| \leqslant r_0. \tag{5.6.7}$$

The inequalities (5.6.6) and (5.6.7) imply that

$$\|u(t)\| \leqslant r(t), \qquad t_0 \leqslant t < \beta, \tag{5.6.8}$$

where $r(t) = r(t, t_0, r_0)$ is the maximal solution of (5.6.4) and (5.6.5). Next we shall establish that $\lim_{t \to \beta-} u(t)$ exists and is a vector in X. For any t_1, t_2 such that $t_0 \leqslant t_1 < t_2 < \beta$ we have

$$\|u(t_1) - u(t_2)\| = \left\| \int_{t_1}^{t_2} f[s, u(s)] \, ds \right\|$$

$$\leqslant \int_{t_1}^{t_2} g(s, \|u(s)\|) \, ds$$

$$\leqslant \int_{t_1}^{t_2} g[s, r(s)] \, ds$$

$$= r(t_2) - r(t_1). \tag{5.6.9}$$

Since $\lim_{t \to \beta_-} r(t)$ exists and is finite, taking limits as $t_1, t_2 \to \beta_-$ and using Cauchy's criterion for convergence, it follows from (5.6.9) that, $\lim_{t \to \beta_-} u(t)$ exists. We now define $u(\beta) \equiv \lim_{t \to \beta_-} u(t)$ and we consider (5.6.1) with $u(\beta)$ as the initial condition at $t = \beta$. In view of the assumed local existence of solution of (5.6.1) through any point in $J \times X$, it follows that $u(t)$ can be extended beyond β, contradicting our assumption. Hence, any solution of (5.6.1) and (5.6.2) exists on $[t_0, \infty)$ and so (5.6.8) and (5.6.9) hold with $\beta = \infty$. Since $r(t)$ is bounded and nondecreasing on J, it follows that $\lim_{t \to \infty} r(t)$ exists and is finite. This and the inequality (5.6.8) and (5.6.9) with $\beta = \infty$ yield the last part of the theorem. The proof is complete.

REMARK 5.6.1. Replacing t by $-t$ a dual of Theorem 5.6.1 can be established for the system

$$du/dt = f(t, u), \qquad t \leqslant t_0, . \tag{5.6.10}$$

$$u(t_0) = u_0 \tag{5.6.11}$$

where $f: I \times X \to X$ and $I = (-\infty, t_0]$. Then under Hypotheses (i) and (ii) of Theorem 5.6.1, with J replaced by I and g of (5.6.4) by $-g$, the conclusion of Theorem 5.6.1 is also true for the solutions of (5.6.10) and (5.6.11) with the $\lim u(t, t_0, u_0)$ now taken as $t \to -\infty$. The intervals J and I above can be replaced by any intervals $[t_0, t_0 + \alpha)$ and $(t_0 - \alpha, t_0]$ respectively. Clearly the hypotheses of this remark are not satisfied by the counterexample of Section 5.1.

DEFINITION 5.6.1. We say that (5.6.1) has *asymptotic equilibrium* if every solution of (5.6.1) through any point $(t_0, u_0) \in J \times X$ tends to a (finite) limit $\xi \in X$ as $t \to \infty$ and conversely to every vector $\xi \in X$ there exists a solution of (5.6.1) which tends to ξ as $t \to \infty$.

When (5.6.1) has asymptotic equilibrium then it is *asymptotically equivalent* to

$$dv/dt = 0 \tag{5.6.12}$$

in the sense that given a solution of (5.6.1) [of (5.6.12)] there exists a solution of (5.6.12) [of (5.6.1)] such that their difference goes to zero as $t \to \infty$.

The next theorem gives a set of sufficient conditions for (5.6.1) to have asymptotic equilibrium.

THEOREM 5.6.2.   Assume that

(i) $f \in C[R_+ \times X, X]$ and maps bounded sets into relatively compact sets;

(ii) $\|f(t, u)\| \leqslant g(t, \|u\|)$, $(t, u) \in R_+ \times X$;

(iii) $g \in C[R_+ \times R_+, R_+]$, $g(t, r)$ is nondecreasing in $r \geqslant 0$ for each $t \in R_+$, and for any $(t_0, r_0) \in R_+ \times R_+$ the maximal solution $r(t, t_0, r_0)$ of (5.6.4) and (5.6.5) is bounded on $[t_0, \infty)$.

Then (5.6.1) has asymptotic equilibrium.

Proof: Let $u(t)$ be a solution of (5.6.1) through (t_0, u_0). By Theorem 5.6.1 $u(t)$ exists on $[t_0, \infty)$ and $\lim_{t \to \infty} u(t)$ exists and is a vector $\xi \in X$. Notice that in proving this part we do not use Hypothesis (i). Conversely, let $\xi \in X$. We must construct a solution $u(t)$ of (5.6.1) which tends to ξ as $t \to \infty$. The proof of this is involved and we shall give it in several steps.

First, as a consequence of Hypothesis (iii) for every $(t_0, \lambda) \in R_+ \times R_+$

$$\int_{t_0}^{\infty} g(s, \lambda) \, ds < \infty. \tag{5.6.13}$$

In fact, let $\tilde{r}(t) = r(t, t_0, \lambda)$ be the maximal solution of (5.6.4) through (t_0, λ). Since $\tilde{r}(t)$ is bounded and nondecreasing the $\lim_{t \to \infty} \tilde{r}(t)$ exists and is a finite number $\tilde{r}_\infty \geqslant \lambda$. From

$$\tilde{r}_\infty \geqslant \tilde{r}(t) = \lambda + \int_{t_0}^{t} g[s, \tilde{r}(s)] \, ds$$

$$\geqslant \lambda + \int_{t_0}^{t} g(s, \lambda) \, ds$$

the condition (5.6.13) follows.

Next, consider the maximal solution $r(t) = r(t, t_0, \|\xi\|)$ of (5.6.4) through $(t_0, \|\xi\|)$. Set $r_\infty = \lim_{t \to \infty} r(t, t_0, \|\xi\|)$. Choose T sufficiently large so that

$$\int_{T}^{\infty} g(s, 2r_\infty) \, ds < r_\infty. \tag{5.6.14}$$

This choice is possible because of (5.6.13).

Now, for each $n = 0, 1, 2, \ldots$ construct the maximal solution $r_n(t) = r(t, T+n, \|\xi\|)$ of (5.6.4) through $(T+n, \|\xi\|)$ and a solution $u_n(t) = u(t, T+n, \xi)$ of (5.6.1) through $(T+n, \xi)$. From Theorem 5.6.1 $u_n(t)$ exists on $[T+n, \infty)$, it tends to a finite limit as $t \to \infty$, and

$$\|u_n(t)\| \leqslant r_n(t), \qquad T+n \leqslant t < \infty. \tag{5.6.15}$$

We shall prove that $u_n(t)$ can be continued backward up to T and that

$$\|u_n(t)\| \leqslant 2r_\infty, \qquad T \leqslant t \leqslant T + n. \tag{5.6.16}$$

Before we do this let $R_n(t) = R_n(t, T+n, \|\xi\|)$ be the maximal solution of the scalar equation

$$r' = -g(t, r) \tag{5.6.17}$$

through $(T+n, \|\xi\|)$. We claim that $R_n(t)$ exists on $[T, T+n]$. To prove this it suffices to show that $R_n(t)$ remains bounded on $[T, T+n]$. If not, there exist points t_1 and t_2 for $T \leqslant t_1 < t_2 \leqslant T+n$ such that $R_n(t_1) = 2r_\infty$ and $R_n(t_2) = r_\infty$ with $r_\infty \leqslant R_n(t) \leqslant 2r_\infty$ on $[t_1, t_2]$. Then from (5.6.17), we have

$$R_n(t_2) = R_n(t_1) - \int_{t_1}^{t_2} g[s, R_n(s)]\, ds.$$

Thus

$$
\begin{aligned}
r_\infty &= \int_{t_1}^{t_2} g[s, R_n(s)]\, ds \\
&\leqslant \int_{t_1}^{t_2} g(s, 2r_\infty)\, ds \\
&\leqslant \int_{T}^{\infty} g(s, 2r_\infty)\, ds
\end{aligned}
$$

contradicting (5.6.14). Now an argument similar to that in the proof of Theorem 5.6.1 (see also Remark 5.6.1) shows that $u_n(t)$ exists on $[T, T+n]$.

Next, we shall establish (5.6.16). If it were false, there should exist points t_3 and t_4 for $T \leqslant t_3 < t_4 \leqslant T+n$ such that $\|u_n(t_3)\| = 2r_\infty$, $\|u_n(t_4)\| = r_\infty$, and $r_\infty \leqslant \|u_n(t)\| \leqslant 2r_\infty$ on $[t_3, t_4]$. Then from (5.6.1) we get

$$u_n(t_4) = u_n(t_3) + \int_{t_3}^{t_4} f[s, u_n(s)]\, ds.$$

Thus

$$
\begin{aligned}
r_\infty &\geqslant 2r_\infty - \int_{t_3}^{t_4} \|f[s, u_n(s)]\|\, ds \\
&\geqslant 2r_\infty - \int_{t_3}^{t_4} g(s, 2r_\infty)\, ds \\
&\geqslant 2r_\infty - \int_{T}^{\infty} g(s, 2r_\infty)\, ds > r_\infty
\end{aligned}
$$

and this contradiction establishes (5.6.16).

The solutions $u_n(t)$ for $n = 0, 1, 2, \ldots$ are therefore defined on $[T, \infty)$ and they are uniformly bounded by $2r_\infty$. Since

$$\|u_n'(t)\| = \|f[t, u_n(t)]\|$$
$$\leqslant g(t, 2r_\infty),$$

the sequence $\{u_n(t)\}$ is equicontinuous on every bounded t interval. We shall now utilize Hypothesis (i) to apply the Ascoli–Arzela Theorem 1.1.1 to this sequence of abstract functions. From (5.6.1) and any fixed $t^* \in [T, \infty)$

$$u_n(t^*) = u_n(T) + \int_T^{t^*} f[s, u_n(s)]\, ds.$$

In view of Hypothesis (i) and Carroll [12, pp. 138, 141–142], it follows that the set of points $\{u_n(t^*)\}$ for $n \geqslant 0$ is relatively compact in X. Thus there is a subsequence, which we still denote by $\{u_n(t)\}$, that converges uniformly on every bounded t interval as $n \to \infty$ to a continuous function $u(t)$.

The function $u(t)$ is the desired solution of (5.6.1) which converges to ξ as $t \to \infty$. In fact

$$u_n(t) = u_n(T) + \int_T^t f[s, u_n(s)]\, ds$$

and passing to the limit we see that $u(t)$ is a solution of (5.6.1). By the first part of the proof of Theorem 5.6.2, $\lim_{t \to \infty} u(t) \equiv u(\infty)$ exists. Since $\lim_{n \to \infty} u_n(t) = u(t)$ and $u_n(T+n) = \xi$ we conclude that $u(\infty) = \xi$. The proof is complete.

REMARK 5.6.2.   The second part of the proof of Theorem 5.6.2 shows under Hypotheses (i), (ii), and (iii) of Theorem 5.6.2 the terminal value Cauchy problem

$$du/dt = f(t, u), \qquad \text{and} \qquad u(\infty) = \xi$$

has a solution for every $\xi \in X$.

PROBLEM 5.6.1.   Establish an asymptotic equilibrium result for (5.6.1) in the case that the function $g(t, u)$ is nonincreasing in u and possibly not defined for $u = 0$.

[Hint: For the case $X = R^n$ see Ladas and Lakshmikantham [38].]

5.7. Lyapunov Functions and Stability Criteria

Here we shall study the stability properties of the solutions of the non-linear Cauchy problem

$$u' = A(t)u + f(t,u) \quad \text{and} \quad u(t_0) = u_0 \in D[A(t_0)] \quad (5.7.1)$$

where $f \in C[R_+ \times X, X]$ and for each $t \in R_+$, $A(t)$ is a linear operator in X with time-varying domain $D[A(t)]$. A solution of (5.7.1) is a strongly differentiable function $u(t)$ such that $u(t) \in D[A(t)]$ for each $t \geq t_0$ and satisfies (5.7.1) for all $t \geq t_0$. We shall assume, without further mention, the existence of solutions $u(t, t_0, u_0)$ of (5.7.1), in the future. We shall also assume that for each $t \in R_+$ and all $h > 0$ but sufficiently small, the operator $R[h, A(t)] \equiv [I - hA(t)]^{-1}$ exists as a bounded operator defined on X, and for each $x \in X$

$$\lim_{h \to 0} R[h, A(t)]x = x. \quad (5.7.2)$$

The following comparison theorem is basic in our discussion of stability criteria.

THEOREM 5.7.1. Assume that

(i) $V \in C[R_+ \times X, R_+]$ and for $(t, x_1), (t, x_2) \in R_+ \times X$

$$|V(t, x_1) - V(t, x_2)| \leq L(t)\|x_1 - x_2\| \quad (5.7.3)$$

where $L(t) \geq 0$ and continuous on R_+;
(ii) there exists a function $g \in C[R_+ \times R_+, R]$ such that for each $(t, x) \in R_+ \times X$

$$D_+ V(t, x) \equiv \lim_{h \to 0_+} \sup h^{-1}[V(t+h, R[h, A(t)]x + hf(t, x)) - V(t, x)]$$

$$\leq g[t, V(t, x)]; \quad (5.7.4)$$

(iii) for each $(t_0, r_0) \in R_+ \times R_+$ the maximal solution $r(t, t_0, r_0)$ of the scalar initial value problem

$$r' = g(t, r) \quad \text{and} \quad r(t_0) = r_0 \quad (5.7.5)$$

exists in the future.

Then $V(t_0, u_0) \leq r_0$ implies that

$$V[t, u(t, t_0, u_0)] \leq r(t, t_0, r_0), \quad t \geq t_0. \quad (5.7.6)$$

Proof: Let $u(t) = u(t, t_0, u_0)$ be any solution of (5.7.1) such that $V(t_0, u_0) \leq r_0$ and $r(t) = r(t, t_0, r_0)$ is the maximal solution of (5.7.5). Define the function $m(t) = V[t, u(t, t_0, u_0)]$. Then

$$m(t_0) \leq r_0. \tag{5.7.7}$$

Further, for $h > 0$ but sufficiently small we obtain, using (5.7.3)

$$m(t+h) - m(t) \leq L(t+h) \|u(t+h) - R[h, A(t)] u(t) - hf[t, u(t)]\|$$
$$+ V[t+h, R(h, A(t)] u(t) + hf[t, u(t)] - V[t, u(t)]. \tag{5.7.8}$$

Since for every $x \in D(A(t))$ we have $R[h, A(t)][I - hA(t)] x = x$, it follows that

$$R[h, A(t)] x + hf(t, x) = x + h[A(t)x + f(t, x)]$$
$$+ h[R(h, A(t)) A(t) x - A(t) x].$$

This together with (5.7.8), implies that

$$m(t+h) - m(t) \leq L(t+h) \|u(t+h) - u(t) - h(A(t)u(t) + f[t, u(t)])\|$$
$$+ L(t+h) h \|R[h, A(t)] A(t) u(t) - A(t) u(t)\|$$
$$+ V(t+h, R[h, A(t)]) u(t) + hf[t, u(t)] - V[t, u(t)].$$

We now use the relations (5.7.1), (5.7.2), and (5.7.4) to obtain

$$D_+ m(t) \leq g[t, m(t)].$$

This and (5.7.7) yields the desired estimate (5.7.6). The proof is complete.

We list a few definitions concerning the stability of the trivial solution of (5.7.1) which we assume to exist for this purpose.

DEFINITION 5.7.1.   The trivial solution of (5.7.1) is said to be

S-1:   equistable if, for each $\varepsilon > 0$ and $t_0 \in R_+$, there exists a positive function $\delta = \delta(t_0, \varepsilon)$ that is continuous in t_0 for each ε such that $\|u_0\| < \delta$ implies $\|u(t, t_0, u_0)\| < \varepsilon$ for $t \geq t_0$;

S-2:   uniformly stable if S-1 holds with δ being independent of t_0;

S-3:   quasi-equi asymptotically stable if, for each $\varepsilon > 0$ and $t_0 \in R_+$, there exist positive numbers $\delta_0 = \delta_0(t_0)$ and $T = T(t_0, \varepsilon)$ such that $\|u_0\| \leq \delta_0$ implies $\|u(t, t_0, u_0)\| < \varepsilon$ for $t \geq t_0 + T$;

S-4: *quasi uniformly asymptotically stable* if S-3 is satisfied with the numbers δ_0 and T independent of t_0;

S-5: *equi-asymptotically stable* if, S-1 and S-3 hold simultaneously;

S-6: *uniformly asymptotically stable* if S-2 and S-4 hold simultaneously.

Let us assume that the scalar equation (5.7.5) possesses the trivial solution also. Then we can define the corresponding stability concepts for the trivial solution of (5.7.5). For example, the trivial solution of (5.7.5) is said to be

S-7': *equistable* if, for each $\varepsilon > 0$ and $t_0 \in R_+$, there exists a positive function $\delta = \delta(t_0, \varepsilon)$ that is continuous in t_0 for each ε such that $r_0 < \delta$ implies $r(t, t_0, r_0) < \varepsilon$ for $t \geq t_0$ where $r(t, t_0, r_0)$ is the maximal solution of (5.7.5). The concepts S-2'–S-6' are defined in a similar way.

We now present a result concerning the equiasymptotic stability of the trivial solution of (5.7.1).

THEOREM 5.7.2. In addition to the hypotheses of Theorem 5.7.1 assume that

(i) $f(t, 0) \equiv 0$, $g(t, 0) \equiv 0$, and $V(t, 0) \equiv 0$, for $t \in R_+$;

(ii) there exists a function $b: R_+ \to R_+$ such that $b(r)$ is increasing in r and

$$b\|x\|) \leq V(t, x), \qquad (t, x) \in R_+ \times X. \tag{5.7.9}$$

Then, S-5' implies S-5. More precisely, S-1' implies S-1 and S-3' implies S-3.

Proof: Suppose that S-1' holds. Let $\varepsilon > 0$ and $t_0 \in R$ be given. Then there exists a $\delta_1 = \delta_1(t_0, \varepsilon) < \varepsilon$, positive and continuous in t_0 for each ε such that $r_0 \leq \delta_1$ implies

$$r(t, t_0, r_0) < b(\varepsilon), \qquad t \geq t_0. \tag{5.7.10}$$

Since $V(t, x)$ is continuous and $V(t, 0) \equiv 0$, there exists a $\delta = \delta(\delta_1, t_0) \leq \delta_1$ such that $\|u_0\| < \delta$ implies $V(t_0, u_0) \leq \delta_1$. This δ depends on t_0 and ε and is continuous in t_0 for each ε. We claim that this δ is good for S-1, that is, $\|u_0\| < \delta$ implies $\|u(t, t_0, u_0)\| < \varepsilon$ for $t \geq t_0$. In fact from the relations (5.7.9), (5.7.6), and (5.7.10) we get

$$b(\|u(t, t_0, u_0)\|) \leq V[t, u(t, t_0, u_0)]$$

$$\leq r(t, t_0, \delta_1)$$

$$< b(\varepsilon), \qquad t \geq t_0.$$

Since $b(r)$ is nondecreasing in r

$$\|u(t, t_0, u_0)\| < \varepsilon, \qquad t \geqslant t_0.$$

Therefore, S-1 holds.

Next we shall prove that S-3′ implies S-3. Let $\varepsilon > 0$ and $t_0 \in R_+$ be given. On the strength of S-3′ there exist positive numbers $\delta_0 = \delta_0(t_0)$ and $T = T[t_0, b(\varepsilon)] = T(t_0, \varepsilon)$ such that $r_0 \leqslant \delta_0$ implies

$$r(t, t_0, r_0) < b(\varepsilon), \qquad t \geqslant t_0 + T. \tag{5.7.11}$$

Since $V(t_0, x)$ is continuous and $V(t_0, 0) = 0$, there exists a $\tilde{\delta}_0 = \tilde{\delta}_0(t_0, \delta_0) < \delta_0$ such that $\|u_0\| < \tilde{\delta}_0$ implies $V(t_0, u_0) < \delta_0$. We claim that $\tilde{\delta}_0$ and T are good for S-3. In fact let $\|u_0\| < \tilde{\delta}_0$ and $t \geqslant t_0 + T$. Then from (5.7.9), (5.7.6), and (5.7.11) we obtain

$$b(\|u(t, t_0, u_0)\|) \leqslant V[t, u(t, t_0, u_0)]$$

$$\leqslant r(t, t_0, \delta_0)$$

$$< b(\varepsilon).$$

Therefore

$$\|u(t, t_0, u_0)\| < \varepsilon, \qquad t \geqslant t_0 + T.$$

The proof is complete.

PROBLEM 5.7.1.   In addition to the hypotheses of Theorem 5.7.1 assume that there exists a function $a: R_+ \to R_+$ such that $a(r)$ is increasing in r and

$$V(t, x) \leqslant a(\|x\|), \qquad (t, x) \in R_+ \times X. \tag{5.7.12}$$

Then S-6′ implies S-6. More precisely, S-2′ implies S-2 and S-4′ implies S-4.

[Hint: Using (5.7.11) the δ's can be chosen independent of t_0.]

It is easy to state and prove various stability and boundedness criteria analogous to the corresponding results in differential equations in Euclidean spaces (see [42]). The main hypotheses in all these results are the existence of a *Lyapunov function* $V(t, x)$ satisfying the hypotheses of Theorem 5.7.1 and, according to the goal, other conditions like (5.7.9), (5.7.11), etc. Since most of the considerations are straightforward, we do not attempt to go into details here.

5.8. Notes

The counterexamples in Section 5.1 are taken from Dieudonné [15] while the counterexample contained in Problem 5.1.1 is due to Yorke [77]. All the results of Section 5.2 are based on the work of Murakami [53]; see also Browder [10]. The nonlinear variations of constants formula and the related material of Section 5.3 are due to Ladas, Ladde, and Lakshmikantham [39]. Some of the ideas here stem from Mamedov [47] and Sultanov [68]. Most of the material presented in Section 5.4 is new and is analogous to the classical work in Lakshmikantham and Leela [42]. See also Brauer [8] and Lumer and Phillips [46]. Section 5.5 consists of the work of Mlak [52]. See Ladas and Lakshmikantham [37] for the material covered in Section 5.6. The proof of the asymptotic equilibrium is fixed rigorously here. Refer to analogous results in Lakshmikantham and Leela [42] for clarification; see also Brauer [7]. For global existence for autonomous differential equations, see Martin [48]. For the stability criteria, using Lyapunov functions, given in Section 5.7 see Lakshmikantham [40] and Lakshmikantham and Leela [42], see also Pao [58], Pao and Vogt [59], Rao and Tsokos [61], and Taam [69]. For further results on the subject the reader is referred to Browder [11].

Chapter 6

Special Topics

6.0. Introduction

As the title of this chapter suggests, here we shall introduce the reader to some topics that are of current interest. In Section 6.1 we present in a simplified way some of the features of nonlinear semigroups and the study of the abstract Cauchy problem

$$du/dt + Au = 0, \quad t \geqslant 0, \quad \text{and} \quad u(0) = u_0$$

where A is a maximal monotone (nonlinear) operator. Section 6.2 introduces the study of delay differential equations in Banach spaces. Since the classical counterpart, namely functional differential equations in finite dimensional spaces, is being investigated at a rapid rate and much has been accomplished in this area we hope that the material of Section 6.2 will induce further study. Here we have presented existence, uniqueness, bounds, continuous dependence, and continuation of solutions of general delay differential equations, in a Banach space, of the form $u' = f(t, u_t)$. Finally in Section 6.3 we discuss second order evolution equations and initiate the study of oscillation theory for such equations in Hilbert spaces.

6.1. Nonlinear Semigroups and Differential Equations

Let X be a real or complex Banach space and S be a subset of X. The notation "lim" (or "w-lim") means the strong limit (or the weak limit) in X.

DEFINITION 6.1.1. A *nonlinear semigroup* in S is a one-parameter family of (possibly nonlinear) operators $\{T(t)\}$, $t \geqslant 0$ from S into itself such that

(i) $\qquad\qquad T(0) = I,\qquad\qquad$ I is the identity on S;

(ii) $\qquad T(t)\,T(s)\,x = T(t+s)\,x,\qquad x \in S,\quad t, s \geqslant 0.$

The semigroup is called *strongly continuous* if for each $x \in S$, $T(t)\,x$ is strongly continuous in $t \geqslant 0$. The semigroup is a *contraction semigroup* if for each $t \geqslant 0$, $T(t)$ is a contraction mapping in S, that is

$$\|T(t)\,x - T(t)\,y\| \leqslant \|x - y\|,\qquad x, y \in S.$$

The *strict infinitesimal generator* A_0 of a nonlinear semigroup $\{T(t)\}$ is defined by

$$A_0\,x = \lim_{h \to 0_+} [T(h)\,x - x]/h$$

and its domain is the set of all $x \in S$ for which the foregoing limit exists in X. The *weak infinitesimal generator* A' of $\{T(t)\}$ is defined by

$$A'x = w - \lim_{h \to 0_+} [T(h)\,x - x]/h$$

and its domain is the set of all $x \in S$ for which this w-limit exists in X.

In this section we shall study the nonlinear abstract Cauchy problem

$$u' + Au = 0 \qquad \text{and} \qquad u(0) = u_0, \tag{6.1.1}$$

where A is a nonlinear operator with domain $D(A) \subset X$ and $u_0 \in D(A)$. The operator A will be assumed to be m-monotonic in the sense of Definition 6.1.4.

DEFINITION 6.1.2.   A function $u: [0, \infty) \to X$ is said to be a

(i)   *strong solution* of (6.1.1) if it is strongly differentiable on $[0, \infty)$ and satisfies (6.1.1) on $[0, \infty)$;

(ii)   *weak solution* of (6.1.1) if it is weakly differentiable on $[0, \infty)$ and satisfies (6.1.1) weakly on $[0, \infty)$;

(iii)   *mild solution* of (6.1.1) if

$$u(t) = u_0 - \int_0^t Au(s)\, ds, \qquad t \geqslant 0,$$

the integral being understood in the Bochner sense.

DEFINITION 6.1.3.   An operator A with domain $D(A)$ and range $R(A)$ in X is said to be *monotonic* if

$$\|x - y + \alpha(Ax - Ay)\| \geqslant \|x - y\|, \qquad x, y \in D(A), \quad \alpha > 0.$$

If A is monotonic, then $(I + \alpha A)^{-1}$ exists for every $\alpha > 0$ and is a contraction operator. Clearly $(I + \alpha A)$ is one to one. Setting $x = (I + \alpha A)^{-1} u$ and $y = (I + \alpha A)^{-1} v$, it follows that

$$\begin{aligned}
\|u - v\| &= \|(I + \alpha A)x - (I + \alpha A)y\| \\
&= \|x - y + \alpha(Ax - Ay)\| \\
&\geqslant \|x - y\| \\
&= \|(I + \alpha A)^{-1}u - (I + \alpha A)^{-1}v\| \qquad (6.1.2)
\end{aligned}$$

and our assertion is established.

DEFINITION 6.1.4.   If A is monotonic and $D(I + \alpha A)^{-1} = X$ for every $\alpha > 0$ then A is called *m-monotonic*.

In the remaining of this section we shall assume that X is a Hilbert space. The following is the main result to be proved:

THEOREM 6.1.1.   Let A on $D(A)$ be m-monotonic in X. Then there exists a unique, strongly continuous semigroup $\{T(t)\}$, $t \geqslant 0$ on $D(A)$ such that for each $u_0 \in D(A)$ we have $u(t) = T(t)u_0$ a locally Lipschitz weak solution of (6.1.1). Furthermore, $T(t)$ is a contraction semigroup having A as weak infinitesimal generator, and for each $u_0 \in D(A)$

(a) $T(t)u_0$ is Lipschitz continuous on $[0, \infty)$;
(b) $T(t)u_0$ is weakly continuous weakly differentiable on $[0, \infty)$;
(c) $T(t)u_0$ is a mild solution of (6.1.1);
(d) $T(t)u_0$ is a strong solution of (6.1.1) except perhaps at a countable number of points;
(e) $T(t)u_0$ is jointly continuous in (t, u_0).

The proof of this theorem will be clear to the reader after a series of sixteen interesting lemmas. The last part (e) is left as an exercise. The key idea behind this proof is to approximate the operator A by a sequence $\{A_n\}$ of everywhere defined monotonic operators defined by $A_n = A(I + n^{-1}A)^{-1}$, and then approximate the solution of (6.1.1) by the sequence $\{u_n(t)\}$ of solutions of the approximating problem

$$u' + A_n u = 0 \quad \text{and} \quad u(0) = u_0 \in D(A).$$

LEMMA 6.1.1. Let $x, y \in X$. Then $\|x\| \leqslant \|x + \alpha y\|$ for every $\alpha > 0$ if and only if $\mathrm{Re}(x, y) \geqslant 0$. In particular A is monotonic if and only if $\mathrm{Re}(Ax - Ay, x - y) \geqslant 0$.

Proof: Let $\mathrm{Re}(x, y) \geqslant 0$. Then

$$\|x + \alpha y\|^2 = \|x\|^2 + 2\alpha \, \mathrm{Re}(x, y) + \alpha^2 \|y\|^2$$

$$\geqslant \|x\|^2.$$

Conversely, let $\|x + \alpha y\|^2 \geqslant \|x\|^2$. Then

$$2\alpha \, \mathrm{Re}(x, y) + \alpha^2 \|y\|^2 \geqslant 0, \qquad \alpha > 0,$$

which implies that $\mathrm{Re}(x, y) \geqslant 0$.

LEMMA 6.1.2. Let $x(t)$ be an X-valued function. If the weak derivative $x'(t)$ exists at $t = s$ and $\|x(t)\|$ is also differentiable at $t = s$, then

$$\|x(s)\| \, \|x(s)\|' = \mathrm{Re}(x'(s), x(s)). \tag{6.1.3}$$

Proof: Since

$$\mathrm{Re}(x(t) - x(s), x(s)) \leqslant (\|x(t)\| - \|x(s)\|) \|x(s)\|,$$

it follows that

$$\mathrm{Re}([x(t) - x(s)]/(t - s), x(s)) \leqslant [(\|x(t)\| - \|x(s)\|)/(t - s)] \|x(s)\|, \quad t > s$$

and

$$\mathrm{Re}([x(t) - x(s)]/(t - s), x(s)) \geqslant [(\|x(t)\| - \|x(s)\|)/(t - s)] \|x(s)\|, \quad t < s.$$

Letting $t \to s$ and using the fact that left and right derivatives are equal, the desired result follows.

For an m-monotonic operator A, we introduce the following sequences of operators:

$$J_n = (I + n^{-1}A)^{-1} \quad \text{and} \quad A_n = AJ_n, \quad n = 1, 2, \dots . \quad (6.1.4)$$

Since the range of J_n is $D(A)$, the operator A_n is well defined. Notice that $A_n = n(I - J_n)$. In fact

$$
\begin{aligned}
A_n &= AJ_n \\
&= n[(I + n^{-1}A) - I](I + n^{-1}A)^{-1} \\
&= n(I - J_n).
\end{aligned}
$$

The operators A_n and J_n are clearly defined everywhere on X.

LEMMA 6.1.3.   Let A be m-monotonic. Then the operators J_n and A_n are Lipschitz continuous, namely they satisfy

$$\|J_n x - J_n y\| \leqslant \|x - y\| \quad \text{and} \quad \|A_n x - A_n y\| \leqslant 2n\|x - y\|. \quad (6.1.5)$$

Proof: The first inequality is a special case of (6.1.2). To prove the second inequality, observe that

$$
\begin{aligned}
\|A_n x - A_n y\| &= n\|x - J_n x - y + J_n y\| \leqslant n\|x - y\| + n\|J_n x - J_n y\| \\
&\leqslant 2n\|x - y\|.
\end{aligned}
$$

LEMMA 6.1.4.   Let A be m-monotonic. Then for each n, A_n is monotonic and

$$\|A_n u\| \leqslant \|Au\|, \quad u \in D(A). \quad (6.1.6)$$

Proof: To prove that A_n is monotonic, in view of Lemma 6.1.1, it suffices to show that $\operatorname{Re}(A_n x - A_n y, x - y) \geqslant 0$. In fact

$$
\begin{aligned}
\operatorname{Re}(A_n x - A_n y, x - y) &= n \operatorname{Re}(x - J_n x - y + J_n y, x - y) \\
&= n[\|x - y\|^2 - \operatorname{Re}(J_n x - J_n y, x - y)] \\
&\geqslant n[\|x - y\|^2 - \|J_n x - J_n y\| \, \|x - y\|] \geqslant 0.
\end{aligned}
$$

Next, from (6.1.5) and for every $u \in D(A)$, we have

$$\|A_n u\| = n \|(I - J_n) u\|$$
$$= n \|J_n (I + n^{-1} A) u - J_n u\|$$
$$\leqslant n \|(I + n^{-1} A) u - u\|$$
$$= \|Au\|$$

and (6.1.6) is proved.

LEMMA 6.1.5. $\lim_{n \to \infty} J_n u = u$ for every $u \in D(A)$.

Proof: From (6.1.6)

$$\|u - J_n n u\| = \|(I - J_n) u\| = n^{-1} \|A_n u\| \leqslant n^{-1} \|Au\| \to 0 \qquad \text{as} \quad n \to \infty.$$

LEMMA 6.1.6. Let A be m-monotonic in X.

(a) If $u_n \in D(A)$ for $n = 1, 2, \ldots$, $\lim_{n \to \infty} u_n = u$, and the $\|Au_n\|$ are bounded for all n, then $u \in D(A)$ and w-$\lim_{n \to \infty} Au_n = Au$;

(b) if $x_n \in X$, $n = 1, 2, \ldots$, $\lim_{n \to \infty} x_n = u \in X$, and the $\|A_n x_n\|$ are bounded for all n, then $u \in D(A)$ and w-$\lim_{n \to \infty} A_n x_n = Au$;

(c) $$\text{w-}\lim_{n \to \infty} A_n u = Au, \qquad u \in D(A).$$

Proof: (a) Since $\|Au_n\|$ is a bounded sequence and X, being a Hilbert space, is reflexive, there exists a subsequence $Au_{n'}$ which converges weakly to a vector $x \in X$ as $n' \to \infty$. Let $v \in D(A)$. Then if $\|Au_n\| \leqslant C$ with $n = 1, 2, \ldots$, we have

$$\text{Re}(Av - Au_{n'}, v - u) = \text{Re}(Av - Au_{n'}, v - u_{n'}) + \text{Re}(Av - Au_{n'}, u_{n'} - u)$$
$$\geqslant \text{Re}(Av - Au_{n'}, u_{n'} - u)$$
$$\geqslant - \|Av - Au_{n'}\| \|u_{n'} - u\|$$
$$\geqslant - (\|Av\| + C) \|u_{n'} - u\|.$$

Taking limits as $n' \to \infty$ we obtain $\text{Re}(Av - x, v - u) \geqslant 0$. Using Lemma 6.1.1 with $\alpha = 1$ we get

$$\|v - u + Av - x\| \geqslant \|v - u\|, \qquad v \in D(A). \tag{6.1.7}$$

Take $v = J_1(u + x) = (I + A)^{-1}(u + x)$ so that $v \in D(A)$ and

$$v + Av = u + x. \tag{6.1.8}$$

The relations (6.1.7) and (6.1.8) imply that $u = v \in D(A)$. From (6.1.8) we then see that $x \equiv$ w-$\lim_{n \to \infty} Au_{n'} = Au$. Since we could have started with any subsequence of $\{u_n\}$ instead of $\{u_n\}$ itself, the foregoing result shows that Au_n converges weakly to Au.

(b) Set $u_n = J_n x_n \in D(A)$. Let $\|A_n x_n\| \leqslant C$ for all n. Then

$$\|Au_n\| = \|AJ_n x_n\|$$

$$= \|A_n x_n\|$$

$$\leqslant C \qquad \text{for all } n.$$

Also

$$\|u - u_n\| = \|u - J_n x_n\|$$

$$= \|u - x_n + n^{-1} A_n x_n\|$$

$$\leqslant \|u - x_n\| + n^{-1} \|A_n x_n\|$$

$$\leqslant \|u - x_n\| + C/n.$$

Therefore, $\lim_{n \to \infty} u_n = u$. The result of (a) is then applicable, proving that $u \in D(A)$ and

$$Au = \text{w-}\lim_{n \to \infty} Au_n$$

$$= \text{w-}\lim_{n \to \infty} Aj_n x_n$$

$$= \text{w-}\lim_{n \to \infty} A_n x_n.$$

(c) Set $x_n = u$. Observe that

$$\|A_n x_n\| = \|A_n u\| \leqslant \|Au\|$$

and therefore (b) is applicable.

The proof is complete.

Since the operator A_n is everywhere defined on X and uniformly Lipschitz continuous, the following result is clear from Theorem 5.1.1.

LEMMA 6.1.7. For each $n = 1, 2, \ldots$, the approximating problem

$$u' + A_n u = 0 \qquad \text{and} \qquad u(0) = u_0 \in D(A) \qquad (6.1.9)$$

has a unique strongly continuously differentiable solution $u_n(t)$ on an interval $[0, \delta_n)$ for some $\delta_n > 0$.

LEMMA 6.1.8. If $u_n(t)$ is the solution of (6.1.9) on $[0, \delta_n)$, then the following estimates hold:

(a) $$\|u_n'(t)\| \leqslant \|Au_0\|, \qquad 0 \leqslant t < \delta_n;$$

(b) $$\|u_n(t+h) - u_n(t)\| \leqslant \|Au_0\| \|h\|, \qquad 0 \leqslant t, t+h < \delta_n.$$

Proof: Since $u_n(t)$ is continuously differentiable on $[0, \delta_n)$, the function $x_n(t) = u_n(t+h) - u_n(t)$ is also continuously differentiable for $0 \leqslant t, t+h < \delta_n$ and hence

$$(d/dt)\|x_n(t)\|^2 = 2\operatorname{Re}(x_n'(t), x_n(t))$$
$$= -2\operatorname{Re}(A_n u_n(t+h) - A_n u_n(t), u_n[t+h] - u_n(t)]) \leqslant 0$$

because A_n is monotonic. Thus $\|x_n(t)\|$ is nonincreasing on $[0, \delta_n)$. In particular

$$\|x_n(t)\|^2 \leqslant \|x_n(0)\|^2.$$

Thus

$$\|u_n(t+h) - u_n(t)\| \leqslant \|u_n(h) - u_n(0)\|.$$

Dividing by $h > 0$ and taking limits as $h \to 0_+$ we obtain

$$\|u_n'(t)\| \leqslant \|u_n'(0)\| = \|A_n u_0\| \leqslant \|Au_0\|$$

which proves (a). The estimate (b) now follows from (a) and the mean value Theorem 1.2.1.

A consequence of Lemmas 6.1.7 and 6.1.8 (b) is the following:

COROLLARY 6.1.1. The approximating problem (6.1.9) has a unique strongly continuously differentiable solution on $[0, \infty)$.

Proof: Clearly it suffices to show that if $u_n(t)$ is a solution of (6.1.9) on $[0, \delta_n)$, then the $\lim_{n \to \delta_{n-}} u_n(t)$ exists. In fact from Lemma 6.1.8

$$\|u_n(t_1) - u_n(t_2)\| \leqslant \|Au_0\| |t_1 - t_2|, \qquad 0 \leqslant t_1, \ t_2 \leqslant \delta_n.$$

Taking limits as $t_1, t_2 \to \delta_{n-}$ and using Cauchy's criterion for convergence our assertion follows.

LEMMA 6.1.9. The sequence $\{u_n(t)\}_{n=1}^{\infty}$ of solutions of (6.1.9) converges strongly to a function $u(t)$ and the convergence is uniform on any finite interval $[0, T]$. Furthermore

$$\|u(t+h) - u(t)\| \leqslant \|Au_0\| \, |h|, \qquad t, t+h \geqslant 0 \qquad (6.1.10)$$

that is, $u(t)$ is Lipschitz continuous on $[0, \infty)$.

Proof: Define $x_{nm}(t) = u_n(t) - u_m(t)$. Then from the monotonicity of A and Lemma 6.1.8(a) we obtain

$$(d/dt)\|x_{nm}(t)\|^2$$

$$= 2 \operatorname{Re}(x'_{nm}(t), x_{nm}(t))$$

$$= -2 \operatorname{Re}(A_n u_n(t) - A_m u_m(t), u_n(t) - u_m(t))$$

$$= -2 \operatorname{Re}(A J_n u_n(t) - A J_m u_m(t), u_n(t) - u_m(t))$$

$$= -2 \operatorname{Re}(A J_n u_n(t) - A J_m u_m(t), J_n u_n(t) - J_m u_m(t))$$

$$\quad - 2 \operatorname{Re}(A_n u_n(t) - A_m u_m(t), u_n(t) - J_n u_n(t) - u_m(t) + J_m u_m(t))$$

$$\leqslant -2 \operatorname{Re}(A_n u_n(t) - A_m u_m(t), n^{-1}A_n u_n(t) - m^{-1}A_m u_m(t))$$

$$\leqslant 2[\|A_n u_n(t)\| + \|A_m u_m(t)\|][\|A_n u_n(t)\|/n + \|A_m u_m(t)\|/m]$$

$$\leqslant 4\|Au_0\|^2(n^{-1} + m^{-1}). \qquad (6.1.11)$$

Integrating (6.1.11) from 0 to t we derive

$$\|x_{nm}(t)\| \leqslant 4\|Au_0\|^2(n^{-1} + m^{-1})t \to 0 \qquad \text{as} \quad n, m \to \infty$$

uniformly for $t \in [0, T]$. The first part of Lemma 6.1.9 is thus established. Now, set $u(t) = \lim_{n \to \infty} u_n(t)$. By Lemma 6.1.8(b)

$$\|u_n(t+h) - u_n(t)\| \leqslant \|Au_0\| \, |h|.$$

Taking limits as $n \to \infty$ the relation (6.1.10) follows. The proof is complete.

LEMMA 6.1.10. $u(t) \in D(A)$ for each $t \geqslant 0$ and w-$\lim_{n \to \infty} A_n u_n(t) = Au(t)$. Furthermore, $Au(t)$ is weakly continuous and $\|Au(t)\| \leqslant \|Au_0\|$.

Proof: By Lemma 6.1.9 we have $\lim_{n \to \infty} u_n(t) = u(t)$ and by Lemma 6.1.8(a) we have $\|A_n u_n(t)\| \leqslant \|Au_0\|$. Now we apply Lemma 6.1.6(b) and conclude that $u(t) \in D(A)$ and w-$\lim_{n \to \infty} A_n u_n(t) = Au(t)$. This means that $\lim_{n \to \infty}(A_n u_n(t), y) = (Au(t), y)$ for each $y \in X$, and

$$|(A_n u_n(t), Au(t))| \leqslant \|A_n u_n(t)\| \, \|Au(t)\|$$

$$\leqslant \|Au_0\| \, \|Au(t)\|, \qquad y = Au(t).$$

Thus

$$|\lim_{n \to \infty} (A_n u_n(t), Au(t))| = \|Au(t)\|^2$$

$$\leqslant \|Au_0\| \|Au(t)\|$$

which implies that

$$\|Au(t)\| \leqslant Au_0\|.$$

To prove the weak continuity of $Au(t)$, it suffices to show that if $t_k \to t$ as $k \to \infty$, then w-$\lim_{k \to \infty} Au(t_k) = Au(t)$. In fact $\lim_{k \to \infty} u(t_k) = u(t)$ and $\|Au(t_k)\| \leqslant \|Au_0\|$. From Lemma 6.1.1 it then follows that w-$\lim_{k \to \infty}$ $Au(t_k) = Au(t)$ and the proof is complete.

LEMMA 6.1.11. $u(t)$ is weakly continous weakly differentiable on $[0, \infty)$ and is a weak solution of (6.1.1).

Proof: By Lemma 6.1.9 $u(0) = \lim_{n \to \infty} u_n(0) = u_0$. Since $u_n(t)$ is a strongly continuously differentiable solution of (6.1.9), we have

$$(u_n(t), v) = (u_0, v) - \int_0^t (A_n u_n(s), v) \, ds, \qquad v \in X. \qquad (6.1.12)$$

By Lemma 6.1.10

$$\lim_{n \to \infty} (A_n u_n(s), v) = (Au(t), v)$$

and by Lemma 6.1.8

$$|(A_n u_n(s), v)| \leqslant \|Au_0\| \|v\|.$$

Thus, the dominated convergence theorem applied to (6.1.12) yields

$$(u(t), v) = (u_0, v) - \int_0^t (Au(s), v) \, ds. \qquad (6.1.13)$$

In view of Lemma 6.1.10 the integrand in (6.1.13) is continuous. Hence $(u(t), v)$ is continuously differentiable and

$$(d/dt)(u(t), v) = -(Au(s), v).$$

The proof is complete.

LEMMA 6.1.12. Let $u(t)$ and $v(t)$ be two Lipschitz continuous weak solutions of $u' + Au = 0$ on $[0, T)$. Then

$$\|u(t) - v(t)\| \leqslant \|u(0) - v(0)\|, \qquad 0 \leqslant t < T. \qquad (6.1.14)$$

In particular, (6.1.1) has exactly one locally Lipschitz weak solution.

Proof: Set $x(t) \equiv u(t) - v(t)$. The Lipschitz character of $u(t)$ and $v(t)$ implies that the function $\|x(t)\|^2$ is absolutely continuous. This fact and Lemma 6.1.2 imply that

$$\|x(t)\|^2 = \|x(0)\|^2 + \int_0^t (d/ds)\|x(s)\|^2 \, ds$$

$$= \|x(0)\|^2 + \int_0^t 2\|x(s)\| \, (d/ds)\|x(s)\| \, ds$$

$$= \|x(0)\|^2 + \int_0^t 2 \operatorname{Re}(x'(s), x(s)) \, ds$$

$$= \|x(0)\|^2 - 2\int_0^t (Au(s) - Av(s), u(s) - v(s)) \, ds$$

$$\leqslant \|x(0)\|^2 ;$$

(6.1.4) follows. In particular, if $u(0) = v(0)$, then $u(t) = v(t)$. The proof is complete.

As a consequence of the parallelogram law in X, or by a direct argument, one can prove the following.

LEMMA 6.1.13.   Let $\{x_n\}$ and $\{y_n\}$ be two sequences of vectors in X such that

$$\lim_{n \to \infty} \|x_n\| = \lim_{n \to \infty} \|y_n\|$$

$$= \tfrac{1}{2} \lim_{n \to \infty} \|x_n + y_n\| < \infty$$

Then

$$\lim \|x_n - y_n\| = 0.$$

LEMMA 6.1.14. $Au(t)$ is strongly continuous except possibly at a countable number of points.

Proof: By Lemma 6.1.10 $\|Au(t)\| \leqslant \|Au(0)\|$. In view of uniqueness one can replace 0 by any $t_0 \leqslant t$ proving that the function $\|Au(t)\|$ is non-increasing in t. Thus $\|Au(t)\|$ is continuous except possibly at a countable number of points. Let $\bar{t}$ be a point of continuity of $\|Au(t)\|$ and let t_k be a sequence of points converging to $\bar{t}$ as $k \to \infty$. Set $x_n = Au(\bar{t})$ and $y_n = Au(t_n)$ for $n = 1, 2, \ldots$. Then

$$\lim_{n \to \infty} \|x_n\| = \lim \|y_n\|$$

$$= \|Au(\bar{t})\|.$$

Since $Au(t)$ is weakly continuous, it also follows that

$$2\|Au(\bar{t})\| \leqslant \lim_{n\to\infty} \inf \|Au(\bar{t}) + Au(t_n)\|$$

$$\leqslant \lim_{n\to\infty} \sup \|Au(\bar{t}) + Au(t_n)\|$$

$$\leqslant \|Au(\bar{t})\| + \lim_{n\to\infty} \|Au(t_n)\|$$

$$= 2\|Au(\bar{t})\|.$$

Thus

$$\lim_{n\to\infty} \|Au(\bar{t}) + Au(t_n)\| = 2\|Au(\bar{t})\|.$$

From Lemma 6.1.13 we then obtain

$$\lim_{n\to\infty} \|Au(\bar{t}) - Au(t_n)\| = 0.$$

Hence, $Au(t)$ is continuous wherever $\|Au(t)\|$ is continuous.

LEMMA 6.1.15. The function $u(t)$ is strongly differentiable except possibly at a countable number of points. Moreover, at the points of strong differentiability $u(t)$ satisfies (6.1.1).

Proof: Let t be a point at which $Au(t)$ is strongly differentiable. Using the weak continuity of $Au(t)$ we notice that

$$\|u(t+h) - u(t) + hAu(t)\|^2$$

$$= (u(t+h) - u(t) + hAu(t), u(t+h) - u(t) + hAu(t))$$

$$= \int_t^{t+h} (-Au(s) + Au(t), u(t+h) - u(t) - hAu(t))\, ds$$

$$= \int_t^{t+h}\int_t^{t+h} (-Au(s) + Au(t), -Au(r) + Au(t))\, dr\, ds$$

$$\leqslant \int_t^{t+h}\int_t^{t+h} \|Au(s) - Au(t)\|\, \|Au(r) - Au(t)\|\, dr\, ds$$

$$\leqslant \varepsilon(h)^2 h^2, \qquad \varepsilon(h) \to 0 \qquad \text{as} \quad h \to 0.$$

Thus

$$\|u(t+h) - u(t)/h + Au(t)\| \leqslant \varepsilon(h) \to 0 \qquad \text{as} \quad h \to 0$$

and the result follows.

LEMMA 6.1.16.   The Lipschitz weak solution of (6.1.1) is also a mild solution of (6.1.1).

Proof: Since $A_n u_n(t)$ is strongly continuous, the values $A_n u_n(t)$ lie in a separable closed linear subspace X_0 of X. Take, for example, X_0 to be the strong closure of the subspace generated by $\{A_n u_n(r)\}$ where r is a rational number and n positive integer. Since w-$\lim_{n\to\infty} A_n u_n(t) = Au(t)$ and X_0 is weakly closed, it follows from Royden [63] that $Au(t) \in X_0$. Thus the values $\{Au(t)\}$ are contained in the closed separable subspace X_0 of X. As $Au(t)$ is w-continuous, it is also w-measurable, and from the results of Section 1.4 $Au(t)$ is strongly measurable. In the proof of Lemma 6.1.14 we have seen that $\|Au(t)\|$ is monotonic and, hence, Lebesgue integrable on every finite interval $[0, T]$. By Theorem 1.4.1 the function $Au(t)$ is Bochner integrable. Since $u(t)$ is a weak solution of (6.1.1), we have using Theorem 1.3.5

$$(u(t), y) = (u_0, y) - \int_0^t (Au(s), y) \, ds$$

$$= (u_0, y) - \left(\int_0^t Au(s) \, ds, y \right)$$

$$= \left(u_0 - \int_0^t Au(s) \, ds, y \right), \qquad y \in X.$$

Hence

$$u(t) = u_0 - \int_0^t Au(s) \, ds.$$

The proof is complete.

PROBLEM 6.1.1.   Complete the proof of Theorem 6.1.1.

[Hint: For $u_0 \in D(A)$ define $T(t)u_0 = u(t)$ where $u(t)$ is the weak solution of (6.1.1).]

PROBLEM 6.1.2.   Let A on $D(A)$ be monotonic in X. Show that $(I+\alpha A)^{-1}$ has domain X either for every $\alpha > 0$ or for no $\alpha > 0$.

[Hint: Observe that $D[(I+\alpha A)^{-1}] = R(I+\alpha A)$ and $R(I+\alpha A) = X$ is equivalent to $R(\alpha^{-1}I+A) = X$.]

PROBLEM 6.1.3.   Let $T(t)x = \max(0, x-t)$ for $x > 0$ and $T(t)x = x$ for $x \leq 0$. Prove that $\{T(t)\}$ is a nonlinear contraction semigroup on R with strict infinitesimal generator A_0 defined by $A_0 x = -1$ for $x > 0$ and $A_0 x = 0$ for $x \leq 0$.

6.2. Functional Differential Equations in Banach Spaces

Suppose $\tau \geqslant 0$ is a given real number and X a Banach space with norm $\|\cdot\|$. Let $\mathscr{C} = C[[-\tau,0], X]$ denote the Banach space of continuous functions mapping the interval $[-\tau,0]$ into X with the norm of $\phi \in \mathscr{C}$ given by

$$\|\phi\|_0 = \max_{-\tau \leqslant s \leqslant 0} \|\phi(s)\|.$$

If $t_0 \in R_+$ and $x \in C[[t_0 - \tau, \infty), X]$, then for any $t \in [t_0, \infty)$, we let $x_t \in \mathscr{C}$ be defined by

$$x_t(s) = x(t+s), \qquad -\tau \leqslant s \leqslant 0.$$

Let $\rho > 0$ be a given constant and let

$$C_\rho = \{\phi \in \mathscr{C} : \|\phi\|_0 \leqslant \rho\}.$$

If $' = d/dt$ and $f: R_+ \times C_\rho \to X$ is a given function, we say that the relation

$$x'(t) = f(t, x_t) \tag{6.2.1}$$

is a *functional differential equation of retarded type* or simply a *functional differential equation.*

DEFINITION 6.2.1.   A function $x(t_0, \phi_0)$ is said to be a *solution* of (6.2.1) with the given initial function $\phi_0 \in C_\rho$ at $t = t_0 \geqslant 0$ if there exists a number $A > 0$ such that

(i) $x(t_0, \phi_0)$ is defined and (strongly) continuous on $[t_0 - \tau, t_0 + A)$ and $x_t(t_0, \phi_0) \in C_\rho$ for $t_0 \leqslant t < t_0 + A$;

(ii) $$x_{t_0}(t_0, \phi_0) = \phi_0;$$

(iii) the (strong) derivative $x'(t_0, \phi_0)$ of $x(t_0, \phi_0)$ at t exists for $t \in [t_0, t_0 + A)$ and satisfies (6.2.1) for $t \in [t_0, t_0 + A)$.

When $\tau = 0$, (6.2.1) reduces to an ordinary differential equation. As we have seen in Section 5.1, mere continuity of f is not enough to guarantee the existence of solutions. We shall present two existence theorems for solutions of (6.2.1). In the first, the function f satisfies a compactness condition, and in the second, f satisfies a Lipschitz condition.

THEOREM 6.2.1.   Let $f \in C[[t_0, t_0 + a] \times C_\rho, X]$. Assume that f maps bounded subsets of $[t_0, t_0 + a] \times C_\rho$ into relatively compact subsets of X.

Then given an initial function ϕ_0 at $t = t_0$, such that $\|\phi_0\|_0 < \rho$, there exists a solution $x(t_0, \phi_0)$ of (6.2.1) on $[t_0 - \tau, t_0 + \alpha]$ where $\alpha > 0$ is sufficiently small.

Proof: Since f maps the bounded set $[t_0, t_0 + a] \times C_\rho$ into a relatively compact set in X, there exists a positive constant M such that $\|f(t, \phi)\| \leqslant M$ for $(t, \phi) \in [t_0, t_0 + a] \times C_\rho$. Choose

$$\alpha = \min\{a, M^{-1}(\rho - \|\phi_0\|_0)\}. \tag{6.2.2}$$

For $n = 1, 2, \ldots$ we define the sequence of functions $\{x^n(t)\}$ by the relation

$$x^n(t) = \begin{cases} \phi_0(-\tau), & t_0 - \tau - 1 \leqslant t \leqslant t_0 - \tau \\ \phi_0(t - t_0), & t_0 - \tau \leqslant t \leqslant t_0 \quad (6.2.3) \\ \phi_0(0) + \displaystyle\int_{t_0}^t f(s, x_{s-n-1}^n)\, ds, & t_0 \leqslant t \leqslant t_0 + \alpha. \end{cases}$$

Clearly $x^n(t)$ is well defined and

$$\|x^n(t)\| < \rho, \qquad t_0 - \tau \leqslant t \leqslant t_0 + \alpha, \qquad (n = 1, 2, \ldots)$$

which implies

$$x_t^n \in C_\rho, \qquad t_0 \leqslant t \leqslant t_0 + \alpha, \quad n = 1, 2, \ldots. \tag{6.2.4}$$

In fact $x^n(t)$ is well defined on $[t_0 - \tau - n^{-1}, t_0 + n^{-1}]$ and in this interval

$$\|x^n(t)\| \leqslant \|\phi_0\|_0 + M(t - t_0)$$
$$\leqslant \|\phi_0\|_0 + M\alpha < \rho.$$

Using again (6.2.3) we can define $x^n(t)$ in $[t_0 + n^{-1}, t_0 + 2n^{-1}]$ and with the same argument as above $\|x^n(t)\| < \rho$ for $t \in [t_0 + n^{-1}, t_0 + 2n^{-1}]$. After a finite number of steps our assertion follows.

From (6.2.3) and (6.2.4) we have, for $t_1, t_2 \in [t_0 - \tau, t_0 + \alpha]$

$$\|x^n(t_1) - x^n(t_2)\| \leqslant M|t_1 - t_2|. \tag{6.2.5}$$

Finally from the compactness of f and Carroll [12] it follows that for each $t^* \in [t_0, t_0 + \alpha]$ the sequence $\{x^n(t^*)\}$ is relatively compact. The last property of the sequence $\{x^n(t)\}$, together with (6.2.4) and (6.2.5) and Ascoli–Arzela Theorem 1.1.1, implies that the sequence $\{x^n(t)\}$ contains a uniformly convergent subsequence which we still denote by $\{x^n(t)\}$. If $x(t)$ is the limit of $\{x^n(t)\}$ as $n \to \infty$, then because of the uniform convergence, $\{x_s^n\}$ also converges to x_s for $-\tau \leqslant s \leqslant 0$. Now taking limits as $n \to \infty$ in (6.2.3), we conclude that $x(t)$ is the desired solution of (6.2.1) on $[t_0 - \tau, t_0 + \alpha]$ with initial function ϕ_0 at $t = t_0$. The proof is complete.

THEOREM 6.2.2. Let $f \in C[[t_0, t_0 + a] \times C_\rho, X]$. Assume that

$$\|f(t, \phi_1) - f(t, \phi_2)\| \leqslant K \|\phi_1 - \phi_2\|_0, \qquad t \in [t_0, t_0 + a], \quad \phi_1, \phi_2 \in C_\rho.$$

Then given an initial function ϕ_0 at $t = t_0$ with $\|\phi_0\|_0 < \rho$ there exists a solution $x(t_0, \phi_0)$ of (6.2.1) on $[t_0 - \tau, t_0 + \alpha]$ where $\alpha > 0$ is sufficiently small.

Proof: We shall employ the contraction mapping principle. In view of the Lipschitz condition there exists a positive constant M such that $\|f(t, \phi)\| \leqslant M$ for $(t, \phi) \in [t_0, t_0 + a] \times C_\rho$. Choose

$$\alpha = \min\{a, M^{-1}(\rho - \|\phi_0\|_0), (2K)^{-1}\}.$$

Let $B = C[[t_0 - \tau, t_0 + \alpha], X]$ be the space of continuous functions x from $[t_0 - \tau, t_0 + \alpha]$ into X such that in addition $x(t) = \phi_0(t - t_0)$ for $t_0 - \tau \leqslant t \leqslant t_0$ and $\|x(t)\| < \rho$ for $t_0 \leqslant t \leqslant t_0 + \alpha$. For $x, y \in B$ we define the distance $d(x, y)$ by

$$d(x, y) = \max_{t_0 \leqslant t \leqslant t_0 + \alpha} \|x(t) - y(t)\|.$$

Clearly B is a complete metric space. For $x \in B$, define

$$Tx(t) = \begin{cases} \phi_0(t), & t_0 - \tau \leqslant t \leqslant t_0 \\ \phi_0(0) + \displaystyle\int_{t_0}^{t} f(s, x_s)\, ds, & t_0 \leqslant t \leqslant t_0 + \alpha. \end{cases} \tag{6.2.6}$$

We have, in view of the choice of α,

$$\|Tx(t)\| \leqslant \|\phi_0\|_0 + M\alpha < \rho.$$

Therefore $T: B \to B$.

Next we shall prove that T is a contraction on B. In fact

$$d(Tx, Ty) \leqslant \max_{t_0 \leqslant t \leqslant t_0 + \alpha} \int_{t_0}^{t} \|f(s, x_s) - f(s, y_s)\|\, ds$$

$$\leqslant K \int_{t_0}^{t} \|x_s - y_s\|\, ds$$

$$\leqslant K\alpha\, d(x, y), \qquad K\alpha \leqslant \tfrac{1}{2}.$$

Hence, the mapping T defined by (6.2.6) has a fixed point x which clearly is a solution of (6.2.1) on $[t_0 - \tau, t_0 + \alpha]$ with initial function ϕ_0 at $t = t_0$. The proof is complete.

We shall next consider a uniqueness theorem of Perron type and also discuss the continuous dependence of solutions with respect to initial values.

THEOREM 6.2.3.   Let $f \in C[[t_0, t_0 + a) \times C_\rho, X]$ and for $t \in [t_0, t_0 + a)$ with $\phi, \psi \in C_\rho$

$$\|f(t, \phi) - f(t, \psi)\| \leqslant g(t, \|\phi - \psi\|_0) \tag{6.2.7}$$

where $g \in C[[t_0, t_0 + a) \times [0, 2\rho), R_+]$. Assume that $r(t) \equiv 0$ is the unique solution of the scalar differential equation

$$r' = g(t, r) \qquad \text{and} \qquad r(t_0) = r_0 \geqslant 0 \tag{6.2.8}$$

with $r_0 = 0$. Then there exists at most one solution of (6.2.1) on $[t_0, t_0 + a)$.

Proof: Suppose that there exist two solutions $x(t_0, \phi_0)$ and $y(t_0, \phi_0)$ of (6.2.1) with the same initial function ϕ_0 at $t = t_0$. Define

$$m(t) = \|x(t_0, \phi_0)(t) - y(t_0, \phi_0)(t)\|$$

so that

$$m_t = \|x_t(t_0, \phi_0) - y_t(t_0, \phi_0)\|.$$

Then, for $t \in (t_0, t_0 + a)$ and using (6.2.7), we obtain

$$D_- m(t) \leqslant \|x'(t_0, \phi_0)(t) - y'(t_0, \phi_0)(t)\|$$
$$= \|f[t, x_t(t_0, \phi_0)] - f[t, y_t(t_0, \phi_0)]\|$$
$$\leqslant g(t, |m_t|_0). \tag{6.2.9}$$

Notice that

$$m_{t_0} = 0. \tag{6.2.10}$$

The inequalities (6.2.9) and (6.2.10) and an application of a known result (Lemma 6.1.1 in Lakshmikantham and Leela [42]) yield

$$\|x(t_0, \phi_0)(t) - y(t_0, \phi_0)(t)\| \leqslant r(t, t_0, 0) \equiv 0, \qquad t \in [t_0, t_0 + a)$$

which implies that

$$x(t_0, \phi_0)(t) \equiv y(t_0, \phi_0), \qquad t \in [t_0, t_0 + a).$$

The proof is complete.

THEOREM 6.2.4. Suppose that the assumptions of Theorem 6.2.3 hold. Assume further the local existence of solutions of (6.2.1). If the solutions $r(t, t_0, r_0)$ of (6.2.8) depend continuously on the initial values (t_0, r_0), then the solutions $x(t_0, \phi_0)$ of (6.2.1) are unique and depend continuously on the initial values (t_0, ϕ_0).

Proof: Since uniqueness follows from Theorem 6.2.3, we prove the continuous dependence. Let $x(t_1, \phi_1)$ and $x(t_2, \phi_2)$ be the solutions of (6.2.1) with initial functions ϕ_1 at $t = t_1$ and ϕ_2 at $t = t_2$, existing in $[t_1, t_0 + \alpha)$ and $[t_2, t_0 + \alpha)$, respectively. Let $t_0 \leqslant t_1 < t_2 < t_0 + a$ and define

$$m(t) = \|x(t_1, \phi_1)(t) - x(t_2, \phi_2)(t)\|, \qquad t \in [t_2, t_0 + \alpha).$$

Then as in Theorem 6.2.3, we arrive at the differential inequality

$$D_- m(t) \leqslant g(t, |m_t|_0), \qquad t \in [t_2, t_0 + \alpha)$$

and

$$|m_{t_2}|_0 = \|x_{t_2}(t_1, \phi_1) - \phi_2\|_0.$$

As before, we then get

$$m(t) \leqslant r(t, t_2, \|x_{t_2}(t_1, \phi_1) - \phi_2\|), \qquad t \in [t_2, t_0 + \alpha) \qquad (6.2.11)$$

where $r(t, t_0, r_0)$ denotes the maximal solution of (6.2.8). Let $(t_2, \phi_2) \to (t_1, \phi_1)$ in the respective topologies. Then since $x_t(t_1, \phi_1)$ is continuous in t, we have $x_{t_2}(t_1, \phi_1) - \phi_2 \to 0$. Using the continuity of $r(t, t_0, r_0)$ in (t_0, r_0), we finally conclude from (6.2.11) that

$$\lim_{\substack{t_2 \to t_1 \\ \phi_2 \to \phi_1}} m(t) = r(t, t_1, 0) \equiv 0.$$

The proof is complete.

PROBLEM 6.2.1. State and prove a uniqueness result parallel to Kamke's uniqueness theorem.

[Hint: Refer to Theorem 6.2.4 in [42].]

DEFINITION 6.2.2. A function $x(t_0, \phi_0, \varepsilon)$ is said to be an ε-approximate solution of (6.2.1) for $t \geqslant t_0$ with an initial function $\phi_0 \in C_\rho$ at $t = t_0$ if

(i) $x(t_0, \phi_0, \varepsilon)$ is defined and strongly continuous on $[t_0 - \tau, \infty)$ and $x_t(t_0, \phi_0, \varepsilon) \in C_\rho$ for $t \geqslant t_0$;

(ii) $$x_{t_0}(t_0, \phi_0, \varepsilon) = \phi_0;$$

(iii) $x(t_0, \phi_0, \varepsilon)$ is (strongly) differentiable on $[t_0, \infty)$, except for an atmost countable set S and satisfies

$$\|x'(t_0, \phi_0, \varepsilon)(t) - f[t, x_t(t_0, \phi_0, \varepsilon)]\| \leqslant \varepsilon, \qquad t \in [t_0, \infty) - S. \quad (6.2.12)$$

In the case where $\varepsilon = 0$ and S is empty, Definition 6.2.2 coincides with Definition 6.2.1.

PROBLEM 6.2.2.   Let $f \in C[R_+ \times C_\rho, X]$ and for (t, ϕ), (t, ψ) ε $R_+ \times C_\rho$ we have

$$\|f(t, \phi) - f(t, \psi)\| \leqslant g(t, \|\phi - \psi\|_0)$$

where $g \in C[R_+ \times [0, 2\rho), R_+]$. Assume that $r(t, t_0, r_0)$ is the maximal solution of

$$r' = g(t, r) + \varepsilon_1 + \varepsilon_2 \qquad \text{and} \qquad r(t_0) = r_0 \geqslant 0.$$

Let $x(t_0, \phi_0, \varepsilon_1)$ and $y(t_0, \psi_0, \varepsilon_2)$ be ε_1- and ε_2-approximate solutions of (6.2.1) such that $\|\phi_0 - \psi_0\| \leqslant r_0$. Then

$$\|x(t_0, \phi_0, \varepsilon_1)(t) - y(t, \psi_0, \varepsilon_2)(t)\| \leqslant r(t, t_0, r_0), \qquad t \geqslant t_0.$$

In the special case, $g(t, r) \equiv Lr$ with $L > 0$, we obtain the well-known estimate

$$\|x(t_0, \phi_0, \varepsilon_1)(t) - y(t_0, \psi_0, \varepsilon_2)(t)\|$$
$$\leqslant \|\phi_0 - \psi_0\|_0 \exp[L(t - t_0)][(\varepsilon_1 + \varepsilon_2)/2][\exp[L(t - t_0)] - 1],$$
$$t \geqslant t_0.$$

Following the proof of Theorem 5.6.1 and the foregoing discussion it is not difficult to prove a global existence result for solutions of (6.2.1) and to show that the (strong) limit of solutions, namely $\lim_{t \to \infty} x(t_0, \phi_0)(t)$, exists and is a vector in X. This we leave as an exercise.

PROBLEM 6.2.3.   Assume that

(i) $f \in C[R_+ \times \mathscr{C}, X]$ and $\|f(t, \phi)\| \leqslant g(t, \|\phi\|_0)$ for all $(t, \phi) \in R_+ \times \mathscr{C}$;
(ii) $g \in C[R_+ \times R_+, R_+]$, $g(t, r)$ is nondecreasing in $r \geqslant 0$ for each $t \in R_+$, and the maximal solution $r(t, t_0, r_0)$ of (6.2.8) exists for $t \geqslant t_0$;
(iii) suppose the local existence of solutions of (6.2.1) through any point $(t_0, \phi_0) \in R_+ \times \mathscr{C}$.

Then the largest interval of existence of any solution $x(t_0, \phi_0)$ of (6.2.1) with $\|\phi_0\|_0 \leqslant r_0$ is $[t_0, \infty)$.

If in addition $r(t, t_0, r_0)$ is assumed to be bounded on $[t_0, \infty)$, then $\lim_{t \to \infty} x(t_0, \phi_0)(t)$ exists and is a (finite) vector in X.

Let us conclude this section with the observation that a number of results concerning bounds, stability, and asymptotic behavior of solutions of functional differential equations in finite dimensional spaces (see Lakshmikantham and Leela [42]) may easily be extended to (6.2.1) in the light of the foregoing discussion.

6.3. Second-Order Evolution Equations

Consider the second-order abstract Cauchy problem

$$u''(t) = B^2 u(t), \qquad t \in R, \tag{6.3.1}$$

$$u(0) = u_1 \qquad \text{and} \qquad u'(0) = u_2, \tag{6.3.2}$$

where B is a linear closed operator with domain $D(B)$ dense in the Banach space X and u_1 and u_2 are given vectors in X.

DEFINITION 6.3.1. A function $u: R \to X$ is said to be a *solution* of (6.3.1) and (6.3.2) if $u \in C_2[R, X]$, $u(t) \in D(B^2)$ for $t \in R$, and satisfies (6.3.1) and (6.3.2).

Define the vector

$$U(t) = \begin{pmatrix} u(t) \\ u'(t) \end{pmatrix} \in X \times X.$$

Then the problem (6.3.1) and (6.3.2) formally becomes

$$U'(t) = MU(t), \qquad t \in R, \tag{6.3.3}$$

$$U(0) = U_0 \tag{6.3.4}$$

where

$$M = \begin{pmatrix} 0 & I \\ B^2 & 0 \end{pmatrix} \qquad \text{and} \qquad U_0 = \begin{pmatrix} u_1 \\ u_2 \end{pmatrix}.$$

If M generates a strongly continuous semigroup (in a sense to be explained later) then the problem (6.3.3) and (6.3.4) has the solutions

$$U(t) = \exp\left\{t\begin{pmatrix} 0 & I \\ B^2 & 0 \end{pmatrix}\right\} U_0.$$

Let us first look at the special case that X coincides with the complex numbers C and B^2 is equal to a constant $b^2 \in C$ with $b \neq 0$. Then

$$\begin{pmatrix} 0 & 1 \\ b^2 & 0 \end{pmatrix}^2 = b^2 \begin{pmatrix} 1 & 0 \\ 0 & 1 \end{pmatrix}$$

and therefore

$$\exp t\begin{pmatrix} 0 & 1 \\ b^2 & 0 \end{pmatrix} = \sum_{n=0}^{\infty} \begin{pmatrix} 0 & 1 \\ b^2 & 0 \end{pmatrix}^n \frac{t^n}{n!}$$

$$= \sum_{n=\text{even}} + \sum_{n=\text{odd}}$$

$$= \cosh(tb)\begin{pmatrix} 1 & 0 \\ 0 & 1 \end{pmatrix} + b^{-1}\sinh(tb)\begin{pmatrix} 0 & 1 \\ b^2 & 0 \end{pmatrix}.$$

The above considerations make plausible the expectation that B generates a strongly continuous group $\{S(t)\}$ on X and if $0 \in \rho(B)$, then M also generates a strongly continuous group $\{T(t)\}$ given by

$$T(t) = \cosh(tB)\begin{pmatrix} I & 0 \\ 0 & I \end{pmatrix} + B^{-1}\sinh(tB)\begin{pmatrix} 0 & 1 \\ B^2 & 0 \end{pmatrix} \qquad (6.3.5)$$

where $\cosh(tB) = \frac{1}{2}[S(t) + S(-t)]$ and $\sinh(tB) = \frac{1}{2}[S(t) - S(-t)]$.

Before we state and prove, with mathematical rigor, the above conjecture, we need some notation. We denote by $[D(B)]^\sim$ the domain of B equipped with the graph norm $|\cdot|$ where $|f| = \|Bf\| + \|f\|$ (or in the case of a Hilbert space $X |f|^2 = \|Bf\|^2 + \|f\|^2$).

With the norm $|\cdot|$, $[D(B)]^\sim$ becomes a Banach (or a Hilbert) space. Notice also that if Z and X are Banach (or Hilbert) spaces, then the space $Y = Z \times X$ is also a Banach (or Hilbert) space provided that it is equipped with the *energy* norm

$$\left\|\begin{pmatrix} f_1 \\ f_2 \end{pmatrix}\right\|_Y = \|f_1\|_Z + \|f_2\|_X.$$

or

$$\left(\left\|\begin{pmatrix} f_1 \\ f_2 \end{pmatrix}\right\|^2_Y = \|f_1\|^2_Z + \|f_2\|^2_X\right).$$

We now prove the following.

THEOREM 6.3.1.   Assume that B with domain $D(B)$ generates a strongly continuous group $\{S(t)\}$ on X and that $0 \in \rho(B)$. Then the operator

$$M = \begin{pmatrix} 0 & I \\ B^2 & 0 \end{pmatrix}$$

with domain $D(M) = D(B^2) \times D(B)$ generates a strongly continuous group $\{T(t)\}$ on $Y = [D(B)]^{\sim} \times X$ given by (6.3.5) and $0 \in \rho(M)$. Furthermore, the Cauchy problem (6.3.1) and (6.3.2) with $u_1 \in D(B^2)$ and $u_2 \in D(B)$ has a unique solution $u(t)$ with the additional property that $u'(t) \in D(B)$ for $t \in R$.

Proof: For

$$g = \begin{pmatrix} g_1 \\ g_2 \end{pmatrix} \in Y,$$

set

$$T(t)g = \begin{pmatrix} \cosh(tB)g_1 + B^{-1}\sinh(tB)g_2 \\ \cosh(tB)g_2 + B\sinh(tB)g_1 \end{pmatrix}.$$

By the definition of Y, it follows that

$$T(t): Y \to Y.$$

By Theorem 2.1.1 (Chapter 2) there exist constants L and ω such that $\|S(t)\| \leq L\exp(\omega|t|)$ for $t \in R$. Then for all $g \in Y$ and $t \in R$

$$\|T(t)g\|_Y = \|\cosh(tB)Bg_1 + \sinh(tB)g_2\|$$

$$+ \|\cosh(tB)g_1 + \sinh(tB)B^{-1}g_2\|$$

$$+ \|\cosh(tB)g_2 + \sinh(tB)Bg_1\|$$

$$\leq L\exp(\omega|t|)(\|Bg_1\| + \|g_2\| + \|g_1\| + \|B^{-1}g_2\| + \|g_2\| + \|Bg_1\|)$$

$$\leq L(2 + \|B^{-1}\|)\exp(\omega|t|)\|g\|_Y.$$

This proves that $T: R \to B(Y)$ and $\|T(\cdot)\|_Y$ is locally bounded. The group property of $\{T\}$ follows easily from the group property of $\{S\}$. Clearly, $T(0) = I$. Next, let $D_0 = D(B^2) \times D(B) \supset Y$. D_0 is dense in Y because $D(B)$ is dense in X and $D(B^2)$ is dense in $[D(B)]^\sim$. To see the last assertion, let $f \in D(B)$. Since $0 \in \rho(B)$, it follows that $B[D(B^2)] = D(B)$ is dense in X. Choose $f_n \in D(B^2)$ such that $\|Bf_n - Bf\| \to 0$ as $n \to \infty$. Then

$$|f_n - f| = \|Bf_n - Bf\| + \|f_n - f\|$$
$$\leqslant (1 + \|B^{-1}\|) \|Bf_n - Bf\| \to 0 \qquad \text{as} \quad n \to \infty.$$

Define

$$N = \begin{pmatrix} 0 & B^{-2} \\ I & 0 \end{pmatrix}.$$

Then

$$\|Ng\|_Y = \|B^{-1}g_2\| + \|B^{-1}g_2\| + \|g_1\|$$
$$\leqslant (1 + \|B^{-1}\|)^2 \|g\|_Y, \qquad g \in Y,$$

and hence $N \in B(Y)$. Notice that $NMf = f$ for all $f \in D(M)$ and $MNf = f$ for all $f \in Y$. Hence $0 \in \rho(M)$ and $M^{-1} = N$.

Next we show that the infinitesimal generator G of $\{T(t)\}$ coincides with M. Clearly $G \supset M$. Assume that $D(M) \not\subseteq D(G)$. Then (since $0 \in \rho(M)$), there exists an $h \in D(G)$ such that $h \neq 0$ and $Gh = 0$. It follows that $T(t)h \equiv h$ for $t \in R$, that is

$$\cosh(tB)h_1 + B^{-1}\sinh(tB)h_2 = h_1, \qquad (6.3.6)$$

$$\sinh(tB)Bh_1 + \cosh(tB)h_2 = h_2. \qquad (6.3.7)$$

Differentiating (6.3.6) we obtain

$$\sinh(tB)Bh_1 + \cosh(tB)h_2 = 0. \qquad (6.3.8)$$

By (6.3.7) and (6.3.8) we get $h_2 = 0$. Then by (6.3.8) $B\sinh(tB)h_1 = 0$ which implies that $\sinh(tB)h_1 = 0$. Taking its derivative at $t = 0$ yields $Bh_1 = 0$, which in turn implies $h_1 = 0$. Hence $h = 0$. This contradiction shows that $G = M$. The remaining part of the theorem follows from the results of Section 2.2. The proof is complete.

Next we shall discuss informally some other simple ways to attack the Cauchy problem (6.3.1) and (6.3.2) as well as some perturbed forms of the same problem.

Assume that B is a closed operator with domain $D(B)$ dense in the Banach space X. Let $u(t)$ be a solution of (6.3.1) and (6.3.2) with the additional property that $u'(t) \in D(B)$ for $t \in R$ and $Bu'(t)$ is continuous in t for $t \in R$. Then

$$(d/dt) Bu(t) = B(d/dt) u(t), \qquad t \in R. \tag{6.3.9}$$

In fact

$$u(t) = u(0) + \int_0^t u'(s) \, ds.$$

By Theorem 1.3.5, it follows that

$$Bu(t) = Bu(0) + \int_0^t Bu'(s) \, ds. \tag{6.3.10}$$

The identity (6.3.9) follows from (6.3.10) and Theorem 1.3.3 (Chapter 1). From (6.3.1) we obtain (by using Theorem 1.3.5)

$$u'(t) = u'(0) + \int_0^t u''(s) \, ds$$

$$= u_2 + B \int_0^t Bu(s) \, ds. \tag{6.3.11}$$

Let v_1 be a vector in X satisfying $u_2 = Bv_1$ (this is possible if, for example, $0 \in \rho(B)$ as in Theorem 6.3.1). Then (6.3.11) becomes

$$u'(t) = B\left[v_1 + \int_0^t Bu(s) \, ds \right]. \tag{6.3.12}$$

Set

$$v(t) = v_1 + \int_0^t Bu(s) \, ds. \tag{6.3.13}$$

From (6.3.12) and (6.3.13) we get

$$u' = Bv \qquad \text{and} \qquad v' = Bu. \tag{6.3.14}$$

Set

$$x = u + v \qquad \text{and} \qquad y = u - v. \tag{6.3.15}$$

Then (6.3.14) yields

$$x' = Bx \qquad \text{and} \qquad y' = -By. \tag{6.3.16}$$

Each of the equations in (6.3.16) can be easily solved if, for example, B is the infinitesimal generator of a strongly continuous group $\{S(t)\}$ (as in Theorem 6.3.1). All the foregoing steps are reversible and therefore we can obtain the solution $u(t) = \frac{1}{2}[x(t) + y(t)]$ of (6.3.1) and (6.3.2) by first solving the simpler equations in (6.3.16). The above arguments give also another proof to the last assertion of Theorem 6.3.1.

Next we consider the more general equation

$$u'' = B(B+Q)u. \tag{6.3.17}$$

Here the equations (6.3.14) will be replaced by

$$u' = Bv \quad \text{and} \quad v' = (B+Q)u.$$

Using the transformations in (6.3.15) we obtain

$$x' = Bx + \tfrac{1}{2}Qx + \tfrac{1}{2}Qy$$

and

$$y' = -By - \tfrac{1}{2}Qx - \tfrac{1}{2}Qy.$$

Introducing the operators

$$\tilde{B} = \begin{pmatrix} B & 0 \\ 0 & -B \end{pmatrix} \quad \text{and} \quad \tilde{S} = \tfrac{1}{2}\begin{pmatrix} Q & Q \\ -Q & -Q \end{pmatrix}$$

in the space $X \times X$ and setting $\tilde{x} = \binom{x}{y}$ we deduce the equation

$$\tilde{x}' = (\tilde{B} + \tilde{S})\tilde{x}$$

which can be treated by the previous methods.

Finally let us examine the nonhomogeneous equation

$$u'' = B^2 u + f(t) \tag{6.3.18}$$

where $f(t)$ is a continuous function. Clearly this equation can be transformed into

$$x' = Bx + B^{-1}f(t), \tag{6.3.19}$$

$$y' = -By - B^{-1}f(t). \tag{6.3.20}$$

Assume, for example, that B generates a continuous group, $0 \in \rho(B)$ and that $B^{-1}f(t)$ is (strongly) continuously differentiable on R. Then Theorem 2.2.3 (Chapter 2) applies to each of the equations (6.3.19) and (6.3.20). In this way, the equation (6.3.18) is also solved.

We shall conclude this chapter by initiating the study of oscillation theory for second-order differential equations of the form

$$[P(t)u'(t)]' + Q(t)u(t) = 0, \qquad t \geqslant 0 \qquad (6.3.21)$$

where $P(t)$ and $Q(t)$ are (generally) unbounded symmetric operators in a Hilbert space H with time-varying domains $D[P(t)]$ and $D[Q(t)]$, respectively. A solution $u(t)$ of (6.3.21) is a continuous function $u: [0, \infty) \to H$ such that for $t \geqslant 0$ we have $u(t) \in D[Q(t)]$, $u'(t) \in D[P(t)]$, $Q(t)u(t)$ continuous, $P(t)u'(t)$ continuously differentiable, and (6.3.21) satisfied. Here we shall assume, without further mention, the existence and uniqueness of solutions of (6.3.21) for any pair of initial conditions $u(t_0) = u_0 \in D[Q(t_0)]$ and $u'(t_0) = u_1 \in D[P(t_0)]$ with $t_0 \geqslant 0$. A function $v: [0, \infty) \to H$ is called nonoscillatory if $v(t)$ vanishes at most once in $[0, \infty)$. We say that the operator A with domain $D(A)$ in H is strictly positive and we write $A > 0$ if and only if $(Ax, x) > 0$ for any $x \in X$ with $x \neq 0$. We write $A < 0$ when $-A > 0$ and $A < B$ when $B - A > 0$.

In the sequel we shall present several nonoscillation results for (6.3.21) and the simpler equation

$$u''(t) + Q(t)u(t) = 0, \qquad t \geqslant 0. \qquad (6.3.22)$$

THEOREM 6.3.2.   Let $P(t) > 0$ and $Q(t) < 0$ for each $t \geqslant 0$. Then every nontrivial solution $u(t)$ of (6.3.21) is nonoscillatory.

Proof: We have

$$0 = ([P(t)u'(t)]' + Q(t)u(t), u(t)) = ([P(t)u'(t)]', u(t)) + (Q(t)u(t), u(t)).$$

Since $Q(t) > 0$ it follows that

$$([P(t)u'(t)]', u(t)) \geqslant 0 \qquad (6.3.23)$$

with strict inequality when $u(t) \neq 0$. Define the function

$$F(t) = (P(t)u'(t), u(t)).$$

Then

$$F'(t) = ([P(t)u'(t)]', u(t)) + (P(t)u'(t), u'(t)) \geqslant 0 \qquad (6.3.24)$$

with strict inequality if $\|u(t)\| + \|u'(t)\| \neq 0$. Since $u(t) \not\equiv 0$, we do have strict inequality in (6.3.23). Hence $F(t)$ is strictly increasing in t. It follows that $F(t)$ has at most one zero in $[0, \infty)$ and therefore $u(t)$ cannot vanish more than once in $[0, \infty)$. This proves that $u(t)$ is nonoscillatory.

PROBLEM 6.3.1.   Let $t^2 Q(t) < \frac{1}{4} I$ for $t > 0$. Then every nontrivial solution $u(t)$ of (6.3.22) is nonoscillatory.

[Hint: Use the transformation $u = t^{1/2} v$ and Theorem 6.3.2.]

PROBLEM 6.3.2.   Let $S(t)$ be a symmetric operator on H such that

$$S'(t) + S^2(t) + Q(t) < 0, \qquad t \geqslant 0.$$

Then every nontrivial solution of (6.3.22) is nonoscillatory.

[Hint: Set $F(t) = (u(t), u'(t)) - (S(t)u(t), u(t))$ and show that $F'(t)$ is strictly increasing.]

PROBLEM 6.3.3.   Prove the converse of Problem 6.3.2 under the additional condition that a solution $U(t)$ of $U'' + Q(t) U = 0$ is invertible for each $t \geqslant 0$.

[Hint: Take $S(t) = U'(t) U^{-1}(t)$.]

PROBLEM 6.3.4.   Let $Q(t)$ be a self-adjoint operator on H and $\{E_\lambda(t)\}_{\lambda = -\infty}^{\lambda = \infty}$ be the resolution of the identity for $Q(t)$. Define the projections

$$E_1(t) = \int_{-\infty}^{0} dE_\lambda(t) \qquad \text{and} \qquad E_2(t) = I - E_1(t).$$

Set $Q_i(t) = E_i(t) Q(t)$ with $i = 1, 2$. Then every nontrivial solution of (6.3.22) is nonoscillatory if one of the following holds:

  (i) $t^2 A_2(t) < -\frac{1}{4} I$;

  (ii)   there exists a symmetric operator $S(t)$ such that

$$S'(t) + S^2(t) + A_2(t) < 0.$$

[Hint: Use Problem 6.3.1 if (i) holds and Problem 6.3.2 when (ii) holds.]

PROBLEM 6.3.5.   Let the symmetric operator Q in (6.3.22) be independent of t and have a nonpositive eigenvalue λ. Then every nontrivial solution of (6.3.22) is nonoscillatory.

[Hint: Let v be the eigenvector of Q corresponding to λ. Set $F(t) = (u(t), v)$ and show that $F'' + \lambda F = 0$.]

PROBLEM 6.3.6.   Let $u(t)$ be a solution of (6.3.22) such that $(u(t_0), u'(t_0))$ is real for some $t_0 \geqslant 0$. Then $(u(t), u'(t))$ is real for each $t \geqslant t_0$.

[Hint: Show that $(u(t), u'(t)) = (u'(t), u(t))$ by showing that their derivatives are equal.]

PROBLEM 6.3.7.   Let $u(t)$ be a solution of (6.3.22) such that $u(t) \neq 0$ for $t \geq t_0$ and $(u(t_0), u'(t_0))$ is real. Assume that $Q(t) \leq q(t)I$ for $t \geq t_0$. Then

$$\|u(t)\|'' + q(t)\|u(t)\| \geq 0.$$

[Hint: Set

$$F(t) = (u'(t), u(t))/(u(t), u(t)), \qquad t \geq t_0$$

and show that $F' + F^2 + q(t) \geq 0$. Notice that

$$\cdot v(t) = \exp \int_{t_0}^{t} F(s) \, ds = \|u(t)\|/\|u(t_0)\|$$

and $v'' + q(t)v \geq 0.$]

6.4. Notes

Section 6.1 presents a simplified version of the results of Kato [32]. For further work in this area the reader is referred to the interesting papers of Dorroh [17], Komura [34], Oharu [57], Mermin [50], Crandall and Liggett [14], Webb [76], Crandall [13], and Brezis and Pazy [9].

The results of Section 6.2 are new in this form. Special cases were previously considered by Mamedov [47] and Zamanov [81]. It remains an open problem to prove an existence theorem for $u' = f(t, u_t)$ similar to Theorem 5.2.1 (Chapter 5) in the text.

Theorem 6.3.1 is due to Goldstein [25]. See Krein [35] and Hille and Phillips [28] with respect to the discussion following the proof of Theorem 6.3.1. For further results and examples see Fattorini [20], Goldstein [26], and Krein [55]. For nonlinear second-order evolution equations refer to Lions and Strauss [45], Strauss [67], and Raskin and Sobolevskii [62]. Higher-order evolution equations are treated in Fattorini [21, 22] and Hille and Phillips [28]. Second-order evolution inequalities are studied in Levine [43] and Agmon [1] where many examples are also given. The nonoscillation results and problems given at the end of Section 6.3 are new.

Appendixes

The purpose of the following appendixes is to give a brief survey of those concepts and results from the theory of functional analysis which are used in the text or frequently used in the related literature. Although these results, for the most part, are standard facts which can be found in any book on the subject, we believe that their inclusion here will help the reader who is not familiar with the subject. We also give references where the proofs of the main results can be found.

Appendix I

Let E be a set. A distance on E is a mapping d of $E \times E$ into the set R of real numbers, having the following properties:

(i) $d(x, y) \geqslant 0, \quad x, y \in E$;

(ii) $d(x, y) = 0 \quad$ if and only if $\quad x = y$;

(iii) $d(x, y) = d(y, x),$ $x, y \in E;$

(iv) $d(x, z) \leqslant d(x, y) + d(y, z),$ $x, y, z \in E$ (triangle inequality).

A *metric space* is a set E together with a given distance d on E.

In a metric space E, a *Cauchy sequence* is a sequence $\{x_n\}$ such that for every $\varepsilon > 0$ there exists a positive integer $N = N(\varepsilon)$ such that $p \geqslant N$ and $q \geqslant N$ imply $d(x_p, x_q) < \varepsilon$.

A metric space E is called *complete* if any Cauchy sequence in E is convergent to a point of E. The importance of complete spaces lies in the fact that to prove that a sequence is convergent in such a space we need only show it is a Cauchy sequence and we do not need to know in advance the value of the limit of the sequence.

A subset D of E is called *dense* in E if $\bar{D} = X$ where $\bar{D}$ denotes the closure of D.

A metric space E is said to be

(i) *separable* if there exists in E an at most denumerable dense set that is, there exists a subset $\{x_1, x_2, \dots\}$ of E which is dense in E;

(ii) *compact* if every open covering of E contains a finite number of sets which is also a covering of E;

(iii) *precompact* (or *totally bounded*) if for every $\varepsilon > 0$ there exists a finite open covering of E by sets of diameter less than or equal to ε (that is, for each $\varepsilon > 0$, E has an ε-*net*).

Clearly if E is precompact, then it is also bounded. Also a compact metric space is separable.

THEOREM (Simmons [69]).   For a metric space E the following conditions are equivalent:

(a) E is compact;

(b)   any infinite sequence of elements of E has a convergent subsequence;

(c) E is precompact and complete.

A subset A of E is said to be *relatively compact* if the closure $\bar{A}$ of A is compact.

THEOREM (Simmons [64]).   A continuous mapping f from a compact metric space E into a metric space E' is uniformly continuous.

Appendix II

Let E be a linear space over the field of real or complex numbers. A *norm* in E is a mapping $x \to \|x\|$ of E into the set R_+ of nonnegative real numbers satisfying the following properties:

(i) $\|x\| \geqslant 0$ and $\|x\| = 0$,   if and only if $x = 0$;

(ii) $\|\lambda x\| = |\lambda| \, \|x\|$, $x \in E$, λ any scalar;

(iii) $\|x+y\| \leqslant \|x\| + \|y\|$, $x, y \in E$ (triangle inequality).

The vector space E together with a norm on E is called a *normed linear space*.

Let us remark here that in a finite dimensional linear space E all norms are equivalent in the sense that if $\|\cdot\|_1$ and $\|\cdot\|_2$ are two norms in E, then there exist nonnegative numbers a and b such that for every $x \in E$

$$a \|x\|_1 \leqslant \|x\|_2 \leqslant b \|x\|_1$$

If E is a normed linear space, then $d(x, y) = \|x-y\|$ is a distance on E. Under this distance, E is a metric space and all the terminology and theorems of metric spaces can also be stated for normed linear spaces.

THEOREM (Goldberg [24]).   If E is a normed space whose unit sphere $S = \{x \in E \colon \|x\| \leqslant 1\}$ is totally bounded, then E is finite dimensional.

A *Banach space* is a normed linear space which is complete (under the distance $d(x, y) = \|x-y\|$).

E is an *algebra* over the scalar field F if E is a linear space over F where multiplication is also defined between the elements of E satisfying the following properties:

(i)   to every ordered pair $x, y \in E$ corresponds a unique element $xy \in E$;

(ii) $(x+y)z = xz + yz$ and $x(y+z) = xy + xz$
                                                (distributivity);

(iii) $(xy)z = x(yz)$ (associativity);

(iv) $\alpha x \beta y = \alpha \beta xy$, $x, y \in E$, $\alpha, \beta \in F$.

A *unit* element in E is a vector $e \in E$ such that $\alpha e = e\alpha$ for every $\alpha \in F$.

A *Banach algebra* over F is a set E which is an algebra as well as a Banach space over F satisfying the additional property $\|xy\| \leqslant \|x\| \, \|y\|$. If a Banach algebra E has a unit element e, then (necessarily $\|e\| \geqslant 1$) we shall assume that $\|e\| = 1$.

Let E be a linear space over the real or complex numbers. An *inner product* on E is a scalar-valued function $(\cdot, \cdot)$ on $E \times E$ with the following properties:

(i) $(\lambda x, y) = \lambda(x, y)$, $\qquad\qquad$ $x, y \in E,$ $\quad$ λ any scalar;

(ii) $(x + y, z) = (x, z) + (y, z)$, $\qquad$ $x, y, z \in E;$

(iii) $(x, y) = \overline{(y, x)}$, $\qquad\qquad\quad$ $x, y \in E;$

(iv) $(x, x) > 0$, $\qquad\qquad\qquad$ $x \neq 0,$ $\quad$ $x \in E.$

A linear space E together with an inner product is called an *inner-product space* (or a *pre-Hilbert space*).

An inner-product space is also a normed space with norm $\|x\| = (x, x)^{1/2}$.

If E is an inner product space the following inequality called *Schwarz's inequality* holds:

$$|(x, y)| \leq \|x\| \|y\|, \qquad x, y \in E$$

with equality if and only if x and y are linearly dependent.

Let X and Y be normed linear spaces. An operator $T: X \to Y$ is called *linear* if $T(\lambda x + \mu y) = \lambda T x + \mu T y$ for every $x, y \in X$ and any scalars λ, μ. The *norm* of the operator T is denoted by $\|T\|$ and is defined by

$$\|T\| = \sup_{\substack{x \in X \\ x \neq 0}} \|Tx\| / \|x\|.$$

One can show that the above definition of the norm of T is also equivalent to

$$\|T\| = \sup_{\substack{\|x\|=1 \\ x \in X}} \|Tx\|$$
$$= \sup_{\substack{\|x\| \leq 1 \\ x \in X}} \|Tx\|.$$

If $\|T\| < \infty$ the T is called a *bounded* operator. Otherwise T is called *unbounded*. Of course every linear transformation T on a finite dimensional space X is bounded.

The set of all bounded operators from X into Y is denoted by $B(X, Y)$. If $X = Y$, we write $B(X)$ instead of $B(X, X)$.

If X is a normed linear space and Y is a Banach space, then $B(X, Y)$ is a Banach space with the norm of $T \in B(X, Y)$ being $\|T\|$. It is easily seen that $B(X)$ is a Banach algebra with identity I (the identity operator) since by the definition of the norm $\|TS\| \leq \|T\| \|S\|$.

THEOREM. If $T: X \to Y$ is a linear operator, then the following statements are equivalent:

(a) T is continuous at the point $x_0 \in X$;
(b) T is uniformly continuous on X;
(c) T is a bounded operator.

A *linear functional* ϕ on X is a linear map from X into the field of scalars (real or complex numbers).

The *conjugate space* X^* of the normed linear space X is the Banach space of all bounded linear functionals on X. The norm of $\phi \in X^*$ is defined by

$$\|\phi\| = \sup_{\substack{x \in X \\ x \neq 0}} |\phi(x)|/\|x\|.$$

A subset K of a linear space X over the real or complex numbers is called *convex* if for every x and y in K, the set $\{\lambda x + (1-\lambda)y : 0 \leqslant \lambda \leqslant 1\}$ is contained in K.

A normed linear space X is called *strictly convex* if $\|x\| = \|y\| = r$ for any r implies $\|x+y\| < 2r$ unless $x = y$ and *uniformly convex* if $\|x_n\| \leqslant 1$, $\|y_n\| \leqslant 1$, and $\|x_n + y_n\| \to 2$ as $n \to \infty$ imply $\|x_n - y_n\| \to 0$ as $n \to \infty$.

A subset M of a linear space X over the complex numbers C is called a *linear manifold* if for every $x, y \in X$ and every $\lambda, \mu \in C$ we have $\lambda x + \mu y \in M$.

A closed linear manifold of X is called a *subspace*.

The adjoint T^* of a linear operator $T \in B(X, Y)$ is the mapping from Y^* to X^* defined by $T^*y^* = y^*T$.

Let T be a linear (not necessarily a bounded) operator with domain $D(T) \supset H$ and range in the Hilbert space H. Assume that $D(T)$ is dense in H. Define $D(T^*) = \{u \in H : \text{ there exists an } f \in H \text{ such that } (u, Tv) = (f, v) \text{ for each } v \in D(T)\}$. Since $\overline{D(T)} = H$ the f associated with u is uniquely determined. Define

$$T^*u = f, \qquad u \in D(T^*).$$

The operator T^* on $D(T^*)$ is called the adjoint of T on $D(T)$.

The operator T on $D(T) \subset H$ is called *symmetric* if $(u, Tv) = (Tu, v)$ for each $u, v \in D(T)$.

The operator T on $D(T) \subset H$ is called *self-adjoint* if

(i) T is symmetric;
(ii) $\overline{D(T)} = H$;
(iii) $T^* = T$.

Let X and Y be normed linear spaces. Suppose that A is a linear operator from X into Y. We say that A is *compact* or *completely continuous* if every bounded set of X is mapped by A into a relatively compact set in Y.

Clearly a compact operator is bounded.

Appendix III

THEOREM (Simmons [64]) (Hahn–Banach). Let M be a subspace of a normed linear space X and ϕ_0 a bounded linear functional on M. Then ϕ_0 can be extended to a linear functional ϕ defined on the whole space X such that $\|\phi\| = \|\phi_0\|$.

Some elementary but very useful consequences of the Hahn–Banach theorem follow.

LEMMA 1. Let $x_0 \neq 0$ be a vector in X. Then there exists a functional $\phi \in X^*$ such that

$$\phi(x_0) = \|x_0\| \qquad \text{and} \qquad \|\phi\| = 1.$$

This lemma implies the existence of many nontrivial bounded functionals on X.

LEMMA 2. If $\phi(x_0) = 0$ for every $\phi \in X^*$, then $x_0 = 0$.

Appendix IV

RIESZ REPRESENTATION THEOREM. Let $X = C[[a,b], R^n]$ be the Banach space of continuous functions from $[a, b]$ into R^n with sup-norm. Let L be a bounded linear functional mapping X into R^n. Then there exists an $n \times n$ matrix $\eta(\theta)$ whose elements are of bounded variation such that for each $\phi \in X$

$$L(\phi) = \int_a^b [d\eta(\theta)]\, \phi(\theta)$$

where the integral is a Stieltjes integral.

Appendix V

Let X be a normed linear space. A linear operator A with domain $D(A) \subset X$ is said to be closed if whenever $x_n \to x$ as $n \to \infty$, $x_n \in D(A)$, and $Ax_n \to y$ as $n \to \infty$, then $x \in D(A)$ and $Ax = y$.

CLOSED-GRAPH THEOREM (Goldberg [24]).   A closed linear operator mapping a Banach space into a Banach space is bounded (and thus continuous).

Appendix VI

Let $\{T_n\}_{n=1}^\infty$, $T_n \in B(X)$ for $n = 1, 2, \ldots$ and $T \in B(X)$. We say that

(i) $T_n \to T$ as $n \to \infty$ in the *strong topology* if $\|T_n x - Tx\| \to 0$ for each $x \in X$ as $n \to \infty$;

(ii) $T_n \to T$ as $n \to \infty$ *in the uniform* operator topology if $\|T_n - T\| \to 0$ as $n \to \infty$;

(iii) $T_n \to T$ as $n \to \infty$ in the *weak topology* if $|\phi(T_n x) - \phi(Tx)| \to 0$ for every $\phi \in X^*$ and each $x \in X$ as $n \to \infty$.

Clearly, uniform convergence implies strong convergence which in turn implies weak convergence.

UNIFORM BOUNDEDNESS PRINCIPLE (Dunford and Schwartz [18]) (Banach–Steinhaus).   Let X and Y be Banach spaces and let $\{T_\alpha\}_{\alpha \in A}$ be an indexed set of bounded linear operators from X into Y. Then the following statements are equivalent:

(a) $\sup_{\alpha \in A} \|T_\alpha\| < \infty$;

(b) $\sup_{\alpha \in A} \|T_\alpha x\| < \infty$, $x \in X$;

(c) $\sup_{\alpha \in A} |\phi(T_\alpha x)| < \infty$, $x \in X$, $\phi \in Y^*$.

The following results are also very useful:

THEOREM (Dunford and Schwartz [18]).   Let X and Y be Banach spaces and let $\{T_\alpha\}_{\alpha \in A}$ be an indexed set of bounded linear operators from X into Y. If for each $x \in X$ the set $\{T_\alpha x : \alpha \in A\}$ is bounded, then $\lim_{x \to 0} T_\alpha x = 0$ uniformly for $\alpha \in A$.

THEOREM (Dunford and Schwartz [18]). Let $T_n: X \to Y$ be a sequence of bounded linear operators from the Banach space X into the Banach space Y. If $\lim_{n \to \infty} T_n x$ exists for each x in a dense subset D of X and if for each $x \in X$ the set $\{T_n x\}$ is bounded, then the limit $Tx \equiv \lim_{n \to \infty} T_n x$ exists for each $x \in X$ and T is a bounded linear operator.

Appendix VII

Let X be a normed linear space and X^{**} the conjugate of the Banach space X^*. The mapping $J_X: X \to X^{**}$ defined by

$$(J_X x)\phi = \phi(x), \qquad \phi \in X^*$$

is called the *natural embedding* of X into X^{**}. If the range of J_X is all of X^{**}, then X is called *reflexive*.

We mention the following facts (see Goldberg [24]):

(i)   the natural embedding $J: X \to X^{**}$ is a linear isometry;
(ii) the conjugate space of a separable reflexive space is separable;
(iii) a Banach space is reflexive if and only if its conjugate is reflexive;
(iv) a closed subspace of a reflexive space is reflexive;
(v)   every bounded sequence in a reflexive space contains a weakly convergent subsequence;
(vi) every Hilbert space is reflexive.

Appendix VIII

Let A be a linear operator (not necessarily bounded) with domain $D(A) \subset X$ and range in the Banach space X. The *resolvent set* of A is the set of all complex numbers λ for which $(\lambda I - A)^{-1}$ exists as a bounded operator with its domain being the whole Banach space X. The resolvent set of A is denoted by $\rho(A)$ and is an open set in the complex plane C. If $\lambda \in \rho(A)$, the function $R(\lambda; A) = (\lambda I - A)^{-1}$ is called the *resolvent function* of A or simply the *resolvent* of A and is an analytic function of $\lambda \in \rho(A)$. *The spectrum* $\sigma(A)$ of A is the complement of $\rho(A)$ in C and therefore is a closed set. If A is a bounded operator, then $\sigma(A)$ is a closed, bounded, and nonempty subset of C. Moreover

$$\sup |\sigma(A)| = \lim_{n \to \infty} (\|A^n\|)^{1/n} \leqslant \|A\|.$$

For $|\lambda| > \sup|\sigma(A)|$ the series

$$R(\lambda; A) = \sum_{n=0}^{\infty} A^n/\lambda^{n+1}$$

converges in the uniform operator topology. The number $r(A) = \sup|\sigma(A)|$ is called the *spectral radius* of A. Let $\sigma(A)$ be the spectrum of the linear operator $A: D(A) \to X$. We define the following:

(i) The *point spectrum* of A, $\sigma_p(A) = \{\lambda \in \sigma(A): \lambda I - A$ is not $1:1\}$. Any point in the point spectrum of A is called an *eigenvalue* of A. If λ is an eigenvalue of A then there exists a vector $x \in D(A)$ for $x \neq 0$ such that $Ax = \lambda x$. The vector x is called an *eigenvector* of A corresponding to the eigenvalue λ.

(ii) The *continuous spectrum* of A, $\sigma_c(A) = \{\lambda \in \sigma(A): \lambda I - A$ is $1:1$ and $(\lambda I - A) D(A)$ is dense in X but not equal to $X\}$.

(iii) The *residual spectrum* of A, $\sigma_r(A) = \{\lambda \in \sigma(A): \lambda I - A$ is $1:1$ and $(\lambda I - A) D(A)$ is not dense in $A\}$.

Clearly $\sigma_p(A), \sigma_c(A)$, and $\sigma_r(A)$ are disjoint and

$$\sigma(A) = \sigma_p(A) \cup \sigma_c(A) \cup \sigma_r(A).$$

If A is a closed operator and λ_0 is a pole of $R(\lambda; A)$ of order m, then λ_0 is an eigenvalue of A. Moreover

$$X = R[(\lambda_0 I - A)^m] \oplus N[(\lambda_0 I - A)^m]$$

where $R[(\lambda_0 I - A)^m]$ and $N[(\lambda_0 I - A)^m]$ are the range and null space respectively, of the operator $(\lambda_0 I - A)^m$.

Let $\lambda, \mu \in \rho(A)$; then we have the identity

$$(\mu I - A)[R(\lambda; A) - R(\mu; A)](\lambda I - A) = (\mu - \lambda) I.$$

Multiplying both sides of this identity by $R(\lambda; A)$ from the right and by $R(\mu; A)$ from the left we get the so called *resolvent formula*

$$R(\lambda; A) - R(\mu; A) = (\mu - \lambda) R(\mu; A) R(\lambda; A).$$

Dividing both sides by $\mu - \lambda$ and letting $\mu \to \lambda$ we obtain

$$(d/d\lambda) R(\lambda; A) = -[R(\lambda; A)]^2.$$

We also remark that for $\lambda, \mu \in \rho(A)$ we have $R_\lambda R_\mu = R_\mu R_\lambda$.

Appendix IX

Let $A: X \to X$ be a bounded operator on the Banach space X. By $F(A)$ we denote the family of all functions f which are analytic on some neighborhood of $\sigma(A)$. The neighborhood need not be connected and can depend on $f \in F(A)$. Let $f \in F(A)$ and let U be an open set whose boundary B consists of a finite number of rectifiable closed Jordan curves, oriented in the positive sense customary in the theory of complex variables. Suppose that $U \supset \sigma(A)$ and that $U \cup B$ is contained in the domain of analyticity of f. Then the operator $f(A)$ is defined by the equation

$$f(A) = (2\pi i)^{-1} \int_B f(\lambda) R(\lambda; A) \, d\lambda.$$

Since $R(\lambda; A)$ is analytic outside $\sigma(A)$ and $f(\lambda)$ is analytic on $U \cup B$, it follows (from Cauchy's integral theorem) that $f(A)$ does not depend on U (but does depend on f).

SPECTRAL MAPPING THEOREM (Dunford and Schwartz [18]). If $f \in F(A)$, then $f[\sigma(A)] = \sigma[f(A)]$.

Appendix X

Let A be a closed operator (in general unbounded) with domain $D(A) \subset X$ and range in the Banach space X. $F(A)$ will denote the family of all functions f which are analytic in some neighborhood of the spectrum $\sigma(A)$ of A and also at infinity. Here we assume that $\rho(A) \neq \varnothing$. For $\alpha \in \rho(A)$ define the operator $T = (A - \alpha I)^{-1} \in B(X)$. Then

$$T(A - \alpha I) x = x, \qquad x \in D(A)$$

and

$$(A - \alpha I) T x = x, \qquad x \in X.$$

We shall define the operational calculus for A in terms of the operational calculus of the bounded operator T. Define the mapping $\Phi: C \to C$ by

$$\phi(\lambda) = (\lambda - \alpha)^{-1} \quad \text{and} \quad \Phi(\alpha) = \infty, \quad \Phi(\infty) = 0.$$

THEOREM (Dunford and Schwartz [18]). If $\alpha \in \rho(A)$, then

$$\Phi[\sigma(A) \cup \{\infty\}] = \sigma(T)$$

and the relation $\phi(\mu) = f[\Phi^{-1}(\mu)]$ determines a one-to-one correspondence between $f \in F(A)$ and $\phi \in F(T)$.

For $f \in F(A)$ we define $f(A) = \phi(T)$ where $\phi \in F(T)$ is given by $\phi(\mu) = f[\Phi^{-1}(\mu)]$.

THEOREM (Dunford and Schwartz [18]). If $f \in F(A)$, then $f(A)$ is independent of the choice of $\alpha \in \rho(A)$. Let U be an open set containing $\sigma(A)$ whose boundary B consists of a finite number of Jordan arcs such that f is analytic on $U \cup B$. Let B have positive orientation with respect to the (possibly unbounded) set U. Then

$$f(A) = f(\infty)I + (2\pi i)^{-1} \int_B f(\lambda) R(\lambda; A) \, d\lambda.$$

Let H be a complex Hilbert space and M be a closed subspace of H. Let $M_\perp$ be the orthogonal complement of M in H, that is, $M_\perp = \{y \in H : (x, y) = 0$ for every $x \in M\}$. It follows that any vector $z \in H$ can be written uniquely as $z = x + y$ where $x \in M$ and $y \in M_\perp$. The mapping $P : H \to H$, defined by taking $Pz = x$ is called the *orthogonal projection* on M. Clearly P is linear, P is idempotent (that is, $P^2 = P$), the range of P is M, and the null space of P is $M_\perp$. It is not difficult to show that a linear operator P is an orthogonal projection if and only if $P^2 = P$ and $P = P^*$.

THEOREM (Resolution of the identity) (see Dunford and Schwartz [19]). Let A with domain $D(A)$ be a self-adjoint operator in the Hilbert space H. Then there exists a family of orthogonal projections $\{E(\lambda)\}$, $\lambda \in R$ such that

(a) $\lambda_1 \leqslant \lambda_2$ implies that $E(\lambda_1) E(\lambda_2) = E(\lambda_2) E(\lambda_1) = E(\lambda_1)$;

(b) $E(\lambda + \varepsilon) \to E(\lambda)$ (strongly)   as $\varepsilon \to 0+$;

(c) $E(\lambda) \to 0$ (strongly)   as $\lambda \to -\infty$,

and

$\qquad\qquad\qquad E(\lambda) \to I$ (strongly)   as $\lambda \to +\infty$;

(d) $$A = \int_{-\infty}^{\infty} \lambda \, dE(\lambda) \qquad\qquad \text{(Stieltjes integral)}$$

and

$$D(A) = \left\{ x \in H : \int_{-\infty}^{\infty} \lambda^2 \, d\|E(\lambda)x\|^2 < \infty \right\}.$$

Bibliography

1. Agmon, S., "Unicité et convexité dans les problèmes differentiels." Presses de L'Univ. de Montréal, Montréal, 1966.
2. Agmon, S., and Nirenberg, L., Properties of solutions of ordinary differential equations in Banach spaces, *Comm. Pure Appl. Math.* **16**, 121–239 (1963).
3. Agmon, S., and Nirenberg, L., Lower bounds and uniqueness theorems for solutions of differential equations in a Hilbert space, *Comm. Pure Appl. Math.* **20**, 207–229 (1967).
4. Alekseev, V. M., An estimate for the perturbations of the solutions of ordinary differential equations (Russian), *Vestnik Moskov. Univ. Ser. I Mat. Meh. No. 2* 28–36 (1961).
5. Aziz, A. K., and Diaz, J. B., On Pompeiu's proof of the mean value theorem of the differential calculus of real-valued functions, *Contrib. Differential Equations* **1**, 467–481 (1963).
6. Boas, R. P., Jr., "A Primer of Real Functions." Mathematical Association of America; distributed by Wiley, New York, 1960.
7. Brauer, F., Global behavior of solutions of ordinary differential equations, *J. Math. Anal. Appl.* **2**, 145–158 (1961).
8. Brauer, F., Perturbations of nonlinear systems of differential equations, I, *J. Math. Anal. Appl.* **14**, 198–206 (1966).

211

9. Brezis, H., and Pazy, A., Accretive sets and differential equations in Banach spaces, *Israel J. Math.* **9**, 367–383 (1970).
10. Browder, F. E., Non-linear equations of evolution, *Ann. of Math.* **80**, 485–523 (1964).
11. Browder, F. E., Nonlinear operators and nonlinear equations of evolution in Banach spaces, *Proc. Pure Math.* **18** (on press).
12. Carroll, R. W., "Abstract Methods in Partial Differential Equations." Harper, New York, 1969.
13. Crandall, M. G., Semigroups of nonlinear transformations in Banach spaces, unpublished manuscript.
14. Crandall, M. G., and Liggett, T. M., Generation of semigroups of nonlinear transformations on general Banach spaces, unpublished manuscript.
15. Dieudonné, J., Deux exemples singuliers d'équations différentielles, *Acta Sci. Math. (Szeged)* **12**, B38–40 (1950).
16. Dieudonné, J., "Foundations of Modern Analysis." Academic Press, New York, 1960.
17. Dorroh, J. R., A nonlinear Hille–Yosida–Phillips theorem, *J. Functional Analysis* **3**, 345–353 (1969).
18. Dunford, N., and Schwartz, J. T., "Linear Operators, Part I: General Theory." Wiley (Interscience), New York, 1958.
19. Dunford, N., and Schwartz, J. T., "Linear Operators, Part II: Spectral Theory." Wiley (Interscience), New York, 1963.
20. Fattorini, H. O., Ordinary differential equations in linear topological spaces, I, *J. Differential Equations* **5**, 72–105 (1969).
21. Fattorini, H. O., On a class of differential equations for vector-valued distributions, *Pacific J. Math.* **32**, 79–104 (1970).
22. Fattorini, H. O., Extension and behavior at infinity of solutions of certain linear operational differential equations, *Pacific J. Math.* **33**, 583–615 (1970).
23. Friedman, A., "Partial Differential Equations." Holt, New York, 1969.
24. Goldberg, S., "Unbounded Linear Operators." McGraw-Hill, New York, 1966.
25. Goldstein, J. A., Semigroups and second-order differential equations, *J. Functional Analysis* **4**, 50–70 (1969).
26. Goldstein, J. A., "Semigroups of Operators and Abstract Cauchy Problems." Lecture notes at Tulane University, 1970.
27. Hale, J., "Functional Differential Equations." Springer-Verlag, Berlin and New York, 1971.
28. Hille, E., and Phillips, R. S., Functional analysis and semi-groups, *Colloq. Amer. Math. Soc.* **31** (1957).
29. Hurd, A. E., Backward lower bounds for solutions of mixed parabolic problems, *Michigan Math. J.* **17**, 97–102 (1970)
30. Kato, T., Integration of the equation of evolution in a Banach space, *J. Math. Soc. Japan* **5**, 208–234 (1953).
31. Kato, T., Nonlinear evolution equations in Banach spaces, *Proc. Symp. Appl. Math.* **17**, 50–67 (1964).
32. Kato, T., Nonlinear semigroups and evolution equations, *J. Math. Soc. Japan* **19**, 508–520 (1967).
33. Kato, T., Semigroups and temporarily inhomogeneous evolution equations, *in* "Equazioni Differenziale Astratte." C.I.M.E. Edizioni Cremonese, Rome, 1963.

34. Komura, Y., Nonlinear semigroups in Hilbert spaces, *J. Math. Soc. Japan* **19**, 493–507 (1967).
35. Krein, S. G., "Linear Differential Equations in a Banach Space" (in Russian). Izdat. Nauka, Moskow, 1967.
36. Ladas, G., and Lakshmikantham, V., Lower bounds and uniqueness for solutions of evolution inequalities in a Hilbert space, *Arch. Rational Mech. Anal.* (on press).
37. Ladas, G., and Lakshmikantham, V., Global existence and asymptotic equilibrium in Banach spaces, *J. Indian Math. Soc.* (on press).
38. Ladas, G., and Lakshmikantham, V., Asymptotic equilibrium of ordinary differential systems, *J. Applicable Anal.* (on press).
39. Ladas, G., Ladde, G., and Lakshmikantham, V., On some fundamental properties of solutions of differential equations in a Banach space, unpublished manuscript.
40. Lakshmikantham, V., Differential equations in Banach spaces and extensions of Lyapunov's method, *Proc. Cambridge Philos. Soc.* **59**, 373–381 (1963).
41. Lakshmikantham, V., Properties of solutions of abstract differential inequalities, *Proc. London Math. Soc.* **14**, 74–82 (1964).
42. Lakshmikantham, V., and Leela, S., "Differential and Integral Inequalities, Theory and Applications," Vols. I and II. Academic Press, New York, 1969.
43. Levine, H. A., Logarithmic Convexity and the Cauchy problem for some abstract second order differential inequalities, *J. Differential Equations* **8**, 34–55 (1970).
44. Lions, J.-L., "Equations différentielles. Opérationelles et problèmes aux limites." Springer Publ., New York, 1961.
45. Lions, J.-L., and Strauss, W. A., Some non-linear evolution equations, *Bull. Soc. Math. France* **93**, 43–96 (1965).
46. Lumer, G., and Phillips, R. S., Dissipative operators in a Banach space, *Pacific J. Math.* **11**, 679–698 (1961).
47. Mamedov, Ya. D., One-sided estimations of the solutions of differential equations with delay argument in Banach spaces, *Trudy Sem. Teor. Differencial. Uravnenii s Otklon. Argumentom Univ. Družby Narodov Patrisa Lumumby* **6**, 135–147 (1968).
48. Martin, R. H., A global existence theorem for autonomous differential equations in a Banach space, *Proc. Amer. Math. Soc.* **26**, 307–314 (1970).
49. Masuda, K., On the exponential decay of solutions for some partial differential equations, *J. Math. Soc. Japan* **19**, 82–90 (1967).
50. Mermin, J. L., An exponential limit formula for nonlinear semigroups, *Trans. Amer. Math. Soc.* **150**, 469–476 (1970).
51. Minty, G. J., Monotone (nonlinear) operators in Hilbert space, *Duke Math. J.* **29**, 541–546 (1962).
52. Mlak, W., Note on abstract differential inequalities and Chaplygin's method, *Ann. Polon. Math.* **10**, 253–271 (1961).
53. Murakami, H., On non-linear ordinary and evolution equations, *Funkcial. Ekvac.* **9**, 151–162 (1966).
54. Ogawa, H., Lower bounds for solutions of differential inequalities in Hilbert space, *Proc. Amer. Math. Soc.* **16**, 1241–1243 (1965).
55. Ogawa, H., On the maximum rate of decay of solutions of parabolic differential inequalities, *Arch. Rational Mech. Anal.* **38**, 173–177 (1970).
56. Ogawa, H., Lower Bounds for solutions of parabolic differential inequalities, *Canad. J. Math.* **19**, 667–672 (1967).

57. Oharu, S., On the generation of semigroups of nonlinear contractions, *J. Math. Soc. Japan* **22**, 526–550 (1970).

58. Pao, C. V., The existence and stability of solutions of nonlinear operator differential equations, *Arch. Rational Mech. Anal.* **35**, 16–29 (1969).

59. Pao, C. V., and Vogt, W. G., On the stability of nonlinear operator differential equations, and applications, *Arch. Rational Mech. Anal.* **35**, 30–46 (1969).

60. Phillips, R. S., Semigroups of contraction operators, *in* "Equazioni Differenziale Astratte." C.I.M.E. Edizioni Cremonese, Rome, 1963.

61. Rao, R. M., and Tsokos, C. P., On the stability and boundedness of differential systems in Banach spaces, *Proc. Cambridge Philos. Soc.* **65**, 507–512 (1969).

62. Raskin, V. G., and Sobolevskii, P. E., The Cauchy problem for second order differential equations in Banach spaces, *Siberian Math. J.* **8**, 52–68 (1967).

63. Royden, H. L., "Real Analysis." Macmillan, New York, 1963.

64. Simmons, C. F.,"Topology and Modern Analysis." McGraw-Hill, New York, 1963.

65. Sobolevskii, P. E., On equations of parabolic type in a Banach space, *Trudy Moscov. Mat. Obšč.* **10**, 297–350 (1961); *Amer. Math. Soc. Transl.* (2), 1–62 (1965).

66. Solomiak, M. Z., The application of the semigroup theory to the study of differential equations in Banach spaces, *Dokl. Akad. Nauk SSSR* **122**, 766–769 (1958).

67. Strauss, W. A., The initial-value problem for certain non-linear evolution equations, *Amer. J. Math.* **89**, 249–259 (1967).

68. Sultanov, R. M., One-sided estimates of solutions of nonlinear differential equations in Banach space, *Siberian Math. J.* **8**, 651–656 (1967).

69. Taam, C. T., Stability, periodicity, and almost periodicity of solutions of nonlinear differential equations in Banach spaces, *J. Math. Mech.* **15**, 849–876 (1966).

70. Tanabe, H., A class of the equations of evolution in a Banach space, *Osaka J. Math.* **11**, 121–145 (1959).

71. Tanabe, H., Remarks on the equations of evolution in a Banach space, *Osaka J. Math.* **12**, 145–166 (1960).

72. Tanabe, H., On the equations of evolution in a Banach space, *Osaka J. Math.* **12**, 363–376 (1960).

73. Tanabe, H., Convergence to a stationary state of the solutions of some kind of differential equations in a Banach space, *Proc. Japan Acad.* **37**, 127–130 (1961).

74. Taylor, A., "Introduction to Functional Analysis." Wiley, New York, 1958.

75. Vainberg, M. M., "Variational Methods for the Study of Nonlinear Operators." Holden-Day, San Francisco, California, 1964.

76. Webb, G. F., Product integral representation of time dependent nonlinear evolution equations in Banach spaces, *Pacific J. Math.* **32**, 269–281 (1970).

77. Yorke, J. A., A continuous differential equation in Hilbert space without existence, *Funkcial. Ekvac.* **13** (1970).

78. Yosida, K., On the differentiability of semigroups of linear operators, *Proc. Japan Acad.* **34**, 337–340 (1958).

79. Zaidman, S., "Equations differentielles abstraites." Presses de L'Univ. de Montréal, Montréal, 1966.

80. Zaidman, S., Some asymptotic theorems for abstract differential equations, *Proc. Amer. Math. Soc.* **25**, 521–525 (1970).

81. Zamanov, T. A., About differential equations with retarded argument in Banach spaces, *Trudy Sem. Teor. Differencial. Uravnenii̇ s Otklon. Argumentom Univ. Družby Narodov Patrisa Lumumby* **4**, 111–115 (1967).

Index

Numbers in italics refer to the pages on which the complete references are listed.

A

Abstract Cauchy problem, 29, 30, 39, 52, 55, 56, 79, 84, 133, 141, 161, 171, 191
Abstract functions, 2
Agmon, S., 125, 199, *211*
Alekseev, V. M., 127, *211*
Approximate solutions, 118
Ascoli–Arzela theorem, 3
Asymptotic behavior, 91, 153
Asymptotic equilibrium, 161, 163
Asymptotic equivalence, 163
Aziz, A. K., 20, *211*

B

Banach algebra, 202
Boas, R. P., Jr., *211*

Bochner integral, *see* Integral
Bounds of solutions, 95, 99, 105, 106, 109, 118, 120, 124, 125
Brezis, H., 199, *212*
Browder, F. E., 171, *212*
Brower, F., 171, *211*

C

Carroll, R. W., 54, 93, 166, 186, *212*
Chain rule for Fréchet derivatives, 19
Chaplygin's method, 127, 157
Closed-graph theorem, 206
Comparison theorem, 167
Conjugate space, 204
Continuous dependence, 22, 133, 145, 175
Counterexamples, 127, 131

215

Convergence (strong, uniform, weak), 206
Convexity results, 95, 101, 103, 104, 106,
   118
Crandall, M. G., 199, *212*

D

Derivative
  Fréchet, 15
  Gateaux, 15
  strong, 2
  weak, 2
Diaz, J. B., 20, *211*
Dieudonné, J., 20, 171, *212*
Differentiability of solutions, 88, 148
Differential
  Fréchet, 15
  Gateaux, 12
Dorroh, J. R., 199, *212*
Dunford, N., 54, 206, 209, 210, *212*

E

Eigenvalue, 208
Equicontinuity, 3
Evolution equations
  first-order linear, 55
  hyperbolic, 57
  nonlinear, 126
  parabolic, 57
  second-order linear, 191
Evolution inequalities, 94
Existence of solutions, 74, 133, 161

F

Fattorini, H. O., 199, *212*
F-differentiable, 16
Fréchet derivative, *see* Derivative
Fréchet differential, *see* Differential
Friedman, A., 54, 93, *212*
Function
  continuous, 2
  countably valued, 8
  Lipschitz continuous, 2
  monotonic, 137
  strongly measurable, 9
  subadditive, 25
  uniformly Hölder continuous, 2
  weakly measurable, 8

Functional differential equations, 42, 185
Fundamental solution
  definition, 56
  existence, 74
  uniqueness, 79

G

Garding's inequality, 49
Gateaux derivative, *see* Derivative
Gateaux differential, *see* Differential
G-differentiable, 15
Goldberg, S., 202, 206, *212*
Goldstein, J. A., 54, 199, *212*
Gradient, 15
Global existence, 161

H

Hale, J., 54, *212*
Hahn–Banach theorem, 205
Heat equation, 41
Hille, E., 20, 22, 32, 54, 199, *212*
Hurd, A. E., 125, *212*

I

Infinitesimal generator, 26, 173
Integral
  Bochner, 9, 10
  improper, 7
  Pettis, 9
  Riemann, 4
  Stieltjes, 10

K

Kato, T., 54, 93, 199, *212*
Komura, Y., 199, *213*
Krein, S. G., 54, 199, *213*

L

Ladas, G., 125, 166, 171, *213*
Ladde, G., 171, *213*
Lakshmikantham, V., 125, 166, 171, 188,
  213
Leela, S., 171, 188, *213*
Levine, H. A., 199, *213*
Liggett, T. M., 199, *212*

Lions, J. L., 125, 199, *213*
Logarithmic norm, 153
Lower bounds of solutions, 95, 106
Lumer, G., 171, *213*
Lyapunov functions, 167, 170

M

Mamedov, Ya. D., 171, 199, *213*
Martin, R. H., 171, *213*
Masuda, K., 125, *213*
Mean value theorem, 3, 26
Mermin, J. L., 199, *213*
Metric space
  compact, 201
  complete, 201
  precompact, 201
  separable, 201
Minty, G. J., 126, *213*
Mlak, W., 171, *213*
m-Monotonic, 174
Monotonic operator, 174
Murakami, H., 171, *213*

N

Nirenberg, L., 125, *211*
Nonoscillation, 197, 198

O

Ogawa, H., 125, *213*
Oharu, S., 199, *214*
Operator
  bounded, 203
  closed, 206
  completely continuous (compact), 205
  monotonic, 174
  m-monotonic, 174
  self-adjoint, 204
  symmetric, 204
  unbounded, 203
Orthogonal projection, 210

P

Pao, C. V., 171, *214*
Parabolic equations, 49, 57, 122, 138
Pazy, A., 199, *212*
Pettis integral, *see* Integral

Phillips, R. S., 20, 22, 32, 54, 171, 199, *212*, *213*, *214*

R

Rao, R. M., 171, *214*
Raskin, V. G., 199, *214*
Reflexive space, 207
Relatively compact, 201
Resolution of the identity, 210
Resolvent formula, 208
Resolvent function, 207
Resolvent set, 33, 207
Riemann integral, *see* Integral
Riesz representation theorem, 205
Royden, H. L., 3, *214*

S

Schwartz, J. T., 54, 206, 209, 210, *212*
Self-adjoint operator, 204
Semigroup
  analytic, 48, 55
  contraction, 23, 173
  nonlinear, 173
  strongly continuous, 23, 173
  uniformly continuous, 23, 173
Simmons, C. F., 201, 205, *214*
Sobolevskii, P. E., 55, 93, 199, *214*
Solomiak, M. Z., 54, 214
Solution (mild, strong, weak), 174
Spectral mapping theorem, 209
Spectral radius, 208
Spectrum
  continuous, 208
  point, 208
  residual, 208
Stability
  asymptotic, 152, 155
  criteria, 167
  equi-, 168
  equi-asymptotic, 169
  quasi-equi asymptotic, 168
  quasi-uniform asymptotic, 169
  uniform asymptotic, 169
  in variation, 153
Strauss, W. A., 199, *213*, *214*
Sultanov, R. M., 171, *214*
Symmetric operator, 204

T

Taam, C. T., 171, *214*
Tanabe, H. A., 55, 93, *214*
Tsokos, C. P., 171, *214*
Taylor, A., 47, *214*

U

Uniform boundedness principle, 206
Uniqueness of solutions, 24, 29, 30, 41, 52,
     79, 95, 106, 118, 133, 139, 140, 144,
     174, 189, 193

V

Vainberg, M. M., 20, *214*

Variation of constants formula, 141
Vogt, W. G., 171, *214*

W

Webb, G. E., 199, *214*

Y

Yorke, J. A., 171, *214*
Yosida, K., 22, 32, 54, *214*

Z

Zaidman, S., 54, 125, *214*
Zamanov, T. A., 199, *214*